浙江省高职院校"十四五"重点立项建设教材

浙江省普通高校"十三五"新形态教材

高等职业教育计算机系列教材

网络故障诊断与排除（H3C）
（第2版）
（微课版）

肖文红　主　编
张厚君　沈晓萍　副主编

电子工业出版社
Publishing House of Electronics Industry
北京·BEIJING

内 容 简 介

本书详细介绍了与 H3C 系列路由器和以太网交换机相关的网络故障分类、网络故障的排除方法、常用的故障诊断工具。本书由 13 个项目组成，内容包括网络管理、网络故障的排除方法与诊断、局域网 VLAN 的故障排除、局域网 STP 的故障排除、交换机 VLAN 间路由故障排除、VRRP 应用的故障排除、静态路由协议应用的故障排除、动态路由协议 RIP 的故障排除、动态路由协议 OSPF 的故障排除、多路由协议共存的故障排除、NAT 网络应用的故障排除、Telnet 协议应用的故障排除和网络自动化运维。

本书落实思政育人，学习贯彻党的二十大精神，将思政元素融入其中，每个项目的开篇设有"素质目标"栏目，每个项目的结尾设有"素质拓展"和"增值服务"栏目，以此增强 IT 信息服务人员的民族自信，提高优质服务意识。

本书取材新颖，内容丰富，重点突出，叙述由浅入深，概念清晰易懂，是一本实用性很强的图书。

本书适合网络管理人员、信息系统管理人员、工程技术人员阅读和参考。

未经许可，不得以任何方式复制或抄袭本书之部分或全部内容。
版权所有，侵权必究。

图书在版编目（CIP）数据

网络故障诊断与排除：H3C：微课版 / 肖文红主编. 2 版. -- 北京：电子工业出版社，2024.8. -- ISBN 978-7-121-48520-6

Ⅰ．TP393.07

中国国家版本馆CIP数据核字第20244UJ537号

责任编辑：徐建军
印　　刷：天津千鹤文化传播有限公司
装　　订：天津千鹤文化传播有限公司
出版发行：电子工业出版社
　　　　　北京市海淀区万寿路 173 信箱　　邮编：100036
开　　本：787×1092　1/16　印张：15.25　字数：372 千字
版　　次：2021 年 9 月第 1 版
　　　　　2024 年 8 月第 2 版
印　　次：2024 年 8 月第 2 次印刷
印　　数：2000 册　定价：54.00 元

凡所购买电子工业出版社图书有缺损问题，请向购买书店调换。若书店售缺，请与本社发行部联系，联系及邮购电话：（010）88254888，88258888。
质量投诉请发邮件至 zlts@phei.com.cn，盗版侵权举报请发邮件至 dbqq@phei.com.cn。
本书咨询联系方式：（010）88254570，xujj@phei.com.cn。

前言

互联网是一个覆盖全球的网络，也是社交、购物、工业等网络的业务数据信息的互动平台，还是未来元宇宙的基础设施，为大众提供各种资源。互联网的主体是终端设备（包括服务器、个人计算机及手持 PC 设备）和网络互联设备（路由器和交换机是其中的重要组成部分）。随着互联网规模的增长，网络管理人员维护网络的任务更加繁重，各种疑难问题也大大增加。因此，快速、有效地解决网络故障问题对提高网络的可用性和可靠性具有非常重要的意义。

本书是浙江省普通高校"十三五"新形态教材《网络故障诊断与排除（H3C）》的修订版。《网络故障诊断与排除（H3C）》（对标国家专业教学标准教材《网络运行与维护》）是高职院校计算机网络相关专业的必修课程。通过对本书的学习，学生可以掌握网络故障的处理思路和处理方法，熟知具体故障的定位和实际处理的步骤，提高处理网络故障的能力。本书由 13 个项目组成，这些项目均是对网络故障诊断与排除的知识总结，涉及 TCP/IP 协议、路由协议、安全管理等方面。本书采用结构化的分析方法，再现了故障处理的整个过程，让学生能够从项目的角度出发，有序地完成问题分析、故障定位和故障恢复等任务，在解决故障的同时巩固网络基础知识。

本书体例新颖，具有以下鲜明特色。

第一，以网络日常运维为主线，在突出计算机网络技术专业的核心技能——网络设备配置与管理的基础上，强化对网络日常运行过程中的管理与维护能力的培养。同时，本书的编写借助了资源库。在教师针对性地引导下，学生可以体验真实的工作场景，学习岗位知识，潜移默化地培养自主学习能力和对工作环境的适应能力。编者在编写本书时考虑了运维工作思路分析和项目文档编写等内容，使得对学生专业核心能力的培养更加贴近企业岗位工作实际，技能养成更加实效。

第二，保持了原有的结构化故障排除方法。本书旨在培养学生在实际故障排除过程中能够以快速分析、准确定位、合理有序的步骤解决网络故障，使学生掌握有效的实践应用方法。

第三，有效地整合了学校教学资源和企业项目资源，着力打造体现生产新工艺、新技术、立体化、自主学习式的新型教材。学生不仅可以依托本书完成传统的课堂学习任务，而且可以在本书中通过图标资源二维码链接所配备的教学资源，方便地开展课后自主拓展学习。本书真正做到了以学生为本，满足学生个体化学习的需求。

第四，增加了"素质目标"和"增值服务"栏目。将基于问题导向的工程思维与严谨细致、不畏困难、精益求精的职业素养相结合，并通过调研客户需求和具体问题，提供合适的解决方案，其实践意义在于引导学生，通过以客户增值体验为中心，以 IT 服务为实现手段，强化增值服务。

本书的网络故障案例均采用 H3C 公司的 H3C Cloud Lab 模拟软件进行系统化再现和处理，易于学生进行系统化学习。

H3C Cloud Lab 是由 H3C 公司发布的一个辅助学习工具，为学习 H3C 网络课程的初学者提供了设计、配置网络和排除网络故障的模拟环境。在这个环境下，学生非常容易实现对故障点的把握，这将有助于他们在未来实际环境中更好地解决故障。

作为学校精品课程及计算机网络技术专业教学资源库的配套教材，本书结合职教云，配有丰富的教学资源，包括课程定位、课程目标及课程内容，电子课件、PPT 演示文稿，原始故障配置的 H3C Cloud Lab 文档（可帮助教师和学生理解产生网络故障的原因和故障现象），排除网络故障后的配置文档（可帮助教师和学生理解解决网络故障的方法和步骤），授课视频文件（提供本书部分内容的教学视频，可供教师和学生进行学习与参考）。

上述教学资源的开发，一方面，可以弥补单一纸质教材的不足，有利于教师利用现代教育技术手段完成教学任务；另一方面，提高了本书的适用性与普及性，支持线上线下同步学习，同时教师利用教学资源并结合本书，可以更好地组织教学活动。

本书由嘉兴职业技术学院的计算机网络技术与信息安全教研团队组织策划，由肖文红担任主编，张厚君、沈晓萍担任副主编。其中，肖文红负责统稿并编写项目 1~项目 3、项目 8~项目 10、项目 12 和项目 13，张厚君负责编写项目 4 和项目 6，沈晓萍负责编写项目 5，谷广兵负责编写项目 7，马焜负责编写项目 11。此外，企业导师陈祖祥（浙江创智科技股份有限公司）参与了本书的编写和指导工作等。

嘉兴职业技术学院的计算机网络技术专业是教育部现代学徒制第二批试点专业。本书是联合浙江创智科技股份有限公司、H3C 公司、杭州洪铭通信技术有限公司等开展现代学徒制试点的重要成果之一，也是校企资源整合、实践资源转化开发的成果。

为了方便教师教学，本书配有电子教学课件及相关教学资源，请对此有需要的教师登录华信教育资源网（www.hxedu.com.cn）下载，如有问题可在网站的留言板中留言或与电子工业出版社联系（E-mail：hxedu@phei.com.cn）。

教材建设是一项系统工程，需要在实践中不断完善和改进，同时由于编者水平有限，书中难免存在疏漏，敬请同行和广大读者给予批评和指正。

编　者

目录

项目1 网络管理 1
1.1 网络管理的基本知识 2
1.1.1 网络管理的概念 2
1.1.2 网络管理的功能 2
1.1.3 网络管理活动 5
1.1.4 网络管理员的任务 6
1.2 网络运行管理制度 7
素质拓展：新理念 新思路 新办法 8
增值服务 9
习题 9

项目2 网络故障的排除方法与诊断 10
2.1 认识网络故障 11
2.2 网络故障的排除方法 11
2.2.1 结构化故障排除的过程 12
2.2.2 故障排除的几种具体方法 14
2.3 网络故障诊断工具 15
2.3.1 故障诊断工具 15
2.3.2 常用的网络测试命令 18
素质拓展：新网络 新科技 新赋能 20
增值服务 21
习题 21

项目3 局域网VLAN的故障排除 22
3.1 VLAN配置分析与实施 23

3.2　VLAN 故障分析与排除 25
3.3　相关知识准备 35
3.4　项目小结 40
素质拓展：测控保驾　神舟飞天 40
增值服务 41
3.5　课后实训 41

项目 4　局域网 STP 的故障排除

4.1　STP 配置分析与实施 43
4.2　STP 配置故障分析与排除 44
4.3　相关知识准备 60
4.4　项目小结 62
素质拓展：安全之道　始于筑基 62
增值服务 63
4.5　课后实训 63

项目 5　交换机 VLAN 间路由故障排除

5.1　VLAN 间路由配置分析与实施 65
5.2　VLAN 间路由配置故障分析与排除 66
5.3　相关知识准备 91
5.4　项目小结 92
素质拓展：炽热青春　开拓创新 92
增值服务 93
5.5　课后实训 93

项目 6　VRRP 应用的故障排除

6.1　VRRP 配置分析与实施 95
6.2　VRRP 配置故障分析与排除 96
6.3　相关知识准备 101
6.4　项目小结 103
素质拓展：网络强国　国之大者 103
增值服务 104
6.5　课后实训 104

项目 7　静态路由协议应用的故障排除

7.1　静态路由配置分析与实施 106

7.2　静态路由配置故障分析与排除 119
　　7.3　相关知识准备 125
　　7.4　项目小结 126
　　素质拓展：弹指一挥　突飞猛进 126
　　增值服务 127
　　7.5　课后实训 127

项目 8　动态路由协议 RIP 的故障排除 128

　　8.1　RIP 配置分析与实施 129
　　8.2　RIP 配置故障分析与排除 139
　　8.3　相关知识准备 147
　　8.4　项目小结 148
　　素质拓展：时不我待　自我创新 148
　　增值服务 149
　　8.5　课后实训 149

项目 9　动态路由协议 OSPF 的故障排除 150

　　9.1　OSPF 配置分析与实施 151
　　9.2　OSPF 配置故障分析与排除 152
　　9.3　相关知识准备 164
　　9.4　项目小结 167
　　素质拓展：移花接木　降本增效 167
　　增值服务 167
　　9.5　课后实训 168

项目 10　多路由协议共存的故障排除 169

　　10.1　多路由协议共存配置分析与实施 170
　　10.2　多路由协议共存配置故障分析与排除 171
　　10.3　相关知识准备 180
　　10.4　项目小结 183
　　素质拓展：雪人计划　利国利民 183
　　增值服务 183
　　10.5　课后实训 183

项目 11　NAT 网络应用的故障排除 185

　　11.1　NAT 配置分析与实施 186

11.2　NAT 配置故障分析与排除 187
　　11.3　相关知识准备 199
　　11.4　项目小结 200
　素质拓展：换道超车　把根留住 200
　增值服务 201
　　11.5　课后实训 202

项目 12　Telnet 协议应用的故障排除 203
　　12.1　Telnet 协议配置分析与实施 204
　　12.2　Telnet 协议配置故障分析与排除 205
　　12.3　相关知识准备 215
　　12.4　项目小结 216
　素质拓展：数字中国　利国利民 216
　增值服务 217
　　12.5　课后实训 217

项目 13　网络自动化运维 218
　　13.1　公司实际需求分析 219
　　13.2　本项目实施具体工作任务 219
　　13.3　项目背景 219
　　　　13.3.1　项目规划设计 220
　　　　13.3.2　项目实施 221
　　　　13.3.3　项目验证 227
　　13.4　项目相关知识 230
　　13.5　项目小结 232
　素质拓展：密码技术　护网安全 232
　增值服务 233
　　13.6　课后实训 233

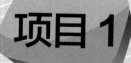

网络管理

内容介绍

在数字化时代,企业和组织都高度依赖网络进行业务运营和信息交流,网络已经成为现代社会的基础设施。网络规模日益扩大,网络环境日益复杂,网络设备种类日益多样,网络问题也变得更加纷繁复杂。专业的技能和经验在网络管理中起着至关重要的作用,可以确保网络设备稳定运行,及时应对各种网络故障,维护企业和组织运营的连续性和数据的安全性。

任务安排

任务1 了解网络管理的基本知识
任务2 掌握网络运行管理制度

学习目标

◇ 了解网络管理的概念
◇ 了解网络管理的功能
◇ 熟记网络管理员的任务
◇ 了解和掌握网络运行管理制度

素质目标

树立坚定的理想信念,实现德、智、体、美、劳全面发展,具有良好的科学文化水平、人文素养、职业道德和创新意识,拥有较强的就业能力和可持续发展能力。

1.1 网络管理的基本知识

随着互联网的发展,各行各业对网络的依赖程度越来越高,各种业务流程和应用软件都在网络系统上运行,这对网络的可用性和可靠性提出了较高的要求。当网络发生故障后,很多业务可能不能如期开展,若不能及时排除故障,则会造成较大损失。因此,这就要求每位网络管理员不仅要配备功能稳定的网络管理软件,还要练就较好的网络故障排除能力,确保能够及时发现网络运行中产生的各种问题,提前解决部分网络隐患,并且能够在网络发生故障后,利用科学的故障排除方法及时、准确地完成故障排除工作,尽量减少因网络故障而造成的损失。

1.1.1 网络管理的概念

网络管理是通过对软/硬件系统的合规使用、人力资源的合理调配、业务流程的优化保障,结合行业经验和企业业务要求,对网络资源进行监控、测试、调度、配置、分析、评价,确保网络系统的稳定、可靠和高性能。网络管理的目标是提供优质的网络运行系统。

1.1.2 网络管理的功能

根据国际标准化组织(ISO)的定义,网络管理有五大功能:故障管理、计费管理、配置管理、性能管理和安全管理。相应地,网络管理软件根据功能的不同,也可分为五类,即网络故障管理软件、网络计费管理软件、网络配置管理软件、网络性能管理软件、网络服务/安全管理软件。就目前网络的实际管理情况来看,故障管理是实际网络运行过程中使用最多的管理模块,故障管理的水平决定着网络运行的整体水平。

网络管理功能

下面具体介绍网络管理的五大功能。

1. 故障管理

当网络运行过程中出现问题时,网络管理员必须做出科学合理的应对方案,有理有据地迅速检查、定位故障并及时排除故障。在通常情况下,发生网络故障的原因比较复杂,一般不能迅速排除,特别是当故障由多个网络组件共同引起时。业内通常采用的故障排除方法是先评估减损,再将网络故障部分进行隔离,继而查找、定位、分析发生网络故障的原因。在分析发生网络故障的原因时尽量做到举一反三,这样做的好处是可对未曾发生的类似故障起到警示预防的作用。故障管理主要包括以下功能。

1)故障监测

主动探测或被动接收网络上的各种事件信息,并识别出其中与网络和系统故障相关的内容,对其中的关键部分保持跟踪,生成网络故障事件日志。

2)故障报警

接收故障监测模块传来的报警信息,根据报警策略驱动不同的报警程序,发出网络故障警报，按设定的提示方式（短信、邮件、电话等）通知网络管理员。

3）故障信息管理

建立日志服务器，收集故障信息，依靠故障记录的分析定位网络故障，记录排除故障的步骤，形成新的网络文档。将"故障信息——故障排除过程——更新后的文档"构成逻辑上相互关联的整体，以反映故障产生的现象、故障定位的思路和故障排除的过程，让故障排除可追溯。

4）提供排错支持工具

网络管理员利用一系列检测方法和检测工具，对发生故障的网络系统进行测试并记录测试结果，以供分析和排错，同时提出排错解决方案。

5）建立故障排错案例库

把每次故障排除过程中生成的各种文档入库，记录典型的故障现象和排错步骤，定期收集故障记录数据，建立故障排错案例库。故障排错案例库要确保能够提供浏览功能，并且能够通过关键字检索故障管理系统中所有的数据记录，为未来的故障排除或远程故障自动诊断和排除积累资源。

2. 计费管理

计费管理一般在具有运营性质的系统中使用较多，主要记录网络资源的使用量，目的是监测和控制网络应用的费用及代价。计费管理主要包括以下功能。

1）计费数据采集

计费数据采集是整个计费系统的基础，受计费网络资源对象和采集使用量的软/硬件设备的制约。

2）数据管理与数据维护

计费管理的数据维护工作可由系统自动完成，但计费依据和范围界定的前期工作，人工交互性很强，需要人为管理，包括计费对象规约、交纳费用的输入、单位信息维护及账单样式的决定等。

3）计费政策制定

由于计费政策经常变化，因此实现用户自由制定计费政策尤其重要。这需要制定一个友好的人机界面和完善地实现计费政策的数据模型。

4）数据查询

每个网络用户可以查询自身使用网络资源的详细信息，根据这些信息可以计算、查询、核对自己的交费情况。

3. 配置管理

网络需要进行一定的配置，以提供相应的服务。配置管理的目的是实现某个特定的网络功能或使网络性能达到最优。配置管理主要包括以下功能。

1）自动获取配置信息

在一个大型网络中，需要管理的设备是比较多的，一个先进的网络管理系统应该具有配置信息自动获取功能，即使在网络管理员不是很熟悉网络结构和配置的情况下，也能通过有关的技术手段来完成对网络的配置和管理，实现网络自动化运维。在较先进的网络管理软件中，都是通过基本模块来实现这一功能的。

2）写配置信息

配置信息自动获取功能相当于从网络设备中"读"信息，相应地，在网络管理中还有大量"写"信息的需求。可以根据设置手段对网络配置信息进行分类：第一类是可以通过网络管理协议标准中定义的方法进行配置的信息；第二类是可以通过自动登录设备进行配置的信息；第三类是需要修改的管理性配置信息。

3）进行配置一致性检查

在一个大型网络中，对网络正常运行影响最大的因素是网络设备接口配置和通信协议配置。由于网络设备众多，因此所有设备很可能不是由同一位网络管理员配置的。实际上，即使是同一位网络管理员对设备进行配置，也会由于各种原因导致配置不一致的问题。因此，对整个网络的配置情况进行一致性检查是必需的，而一致性检查主要就是检查接口配置和通信协议配置。

4）记录用户操作

配置系统的安全性是整个网络管理系统安全的核心，因此必须记录并保存用户的每一次配置操作。网络管理员可以随时查看特定用户在特定时间内的配置操作，实现操作的跟踪和溯源。

4．性能管理

性能管理是对系统资源的运行状况及通信效率等系统性能的管理，其主要操作是监视和分析网络及其所提供服务的性能，性能分析的结果用来确定是否需要优化网络。性能管理主要包括以下功能。

1）性能监控

由用户定义被管对象及其属性，被管对象包括服务器、交换机和路由器，其属性包括流量、延迟、丢包率、CPU 利用率、温度、内存余量等。对于每个被管对象，定时采集其性能数据，自动生成性能报告。

2）可视化的性能报告

对数据进行加工处理，生成性能趋势曲线，以直观的图形反映性能分析的结果。

3）性能分析

统计、整理并分析历史数据，计算性能指标，对性能状况做出判断，为网络优化提供参考。

4）网络对象性能查询

可通过列表或按关键字查询被管网络对象及其属性的性能记录。

5）阈值控制

可设置每个被管对象的每条属性的阈值，对于特定被管对象的特定属性，可以针对不同的时间段和性能指标进行阈值设置。通过阈值检查开关可以控制阈值报警。

5. 安全管理

目前，网络中主要有以下几大安全问题：网络数据的私有性、完整性，合理授权，访问控制等。网络安全管理由以下机制来保证。

① 管理员（网络管理系统的用户）身份认证采用基于公开密钥的证书认证机制。为提高系统效率，在信任域（如局域网）内的用户可以使用口令认证。

② 管理信息存储和传输的加密与完整性。内部存储信息采用加密机制，Web 浏览器和网络管理服务器之间采用安全套接字层（SSL）传输协议，加密传输并保证其完整性。

③ 管理员的分组管理与访问控制。管理员按任务的不同可以分成若干用户组，不同的用户组有不同的权限范围，受访问控制约束检查。

④ 系统日志分析。日志可以记录用户所有的操作，使系统的操作和对网络对象的修改有据可查，有助于故障的跟踪与恢复。

1.1.3 网络管理活动

目前，网络管理的主要内容是网络运维，网络运维工作以服务为中心，以稳定、安全、高效为基本点，确保公司网络业务能够 7×24 小时为用户提供高质量的服务。运维人员需要对公司网络业务所依赖的基础设施、基础服务、线上业务进行稳定性保障和日常巡检，发现服务可能存在的隐患，对整体架构进行优化，以屏蔽常见的运行故障。

网络管理活动是指为保障网络与业务正常、安全、有效运行而采取的生产组织管理活动，简称运维管理或 OAM（Operation Administration and Maintenance），负责维护并确保整个服务的高可用性，同时不断优化系统架构，提升部署效率。

因此，网络管理活动的工作应包括以下 3 类。

Operation：配置操作，包括登录管理、用户管理和业务配置操作等。

Administration：资源管理，包括硬件资源管理和软件资源管理。

Maintenance：维护，包括例行维护和故障维护。

配置操作主要是指日常网络和业务进行的分析、规划和配置工作，主要包括登录配置、用户注册和授权、认证方式、数据转发、路由编排、安全策略和系统优化等。

硬件资源管理主要包括备份电子标签、配置 CPU 占用率告警阈值、配置内存占用率告警阈值、单板管理、光模块告警管理、节能管理等。

软件资源管理主要包括 License 管理、系统管理、接口管理等。

网络管理主要指维护，包括例行维护和故障维护。例行维护的目的是通过日常的例行维护发现并消除设备的运行隐患，主要包括设备环境检查、设备基本信息检查、设备运行状态检查、接口内容检查、业务检查等。

故障维护是针对网络故障进行的修复和恢复工作，当网络设备发生故障时，需要及时排查问题、诊断和修复故障，以恢复网络的正常运行。遵循的基本步骤是观察现象、收集信息、判断分析、原因排查，一般可以分 3 个阶段：故障信息采集阶段、故障定位与诊断阶段、故障修复阶段。

运维人员在进行网络维护时的注意事项如下。

①在发生故障时，请先评估是否为紧急故障，若是紧急故障，则使用预先制定的紧急故障排除方法尽快恢复故障模块，以恢复业务。

②严格遵守操作规程和行业安全规程，确保人身安全与设备安全。

③在更换和维护设备部件的过程中，要做好防静电措施，佩戴防静电腕带。

④在故障排除的过程中遇到任何问题时，应详细记录各种原始信息。

⑤所有的重大操作，如重启设备、擦除数据库等均应做记录，并在操作前仔细确认操作的可行性，在做好相应的风险评估、操作计划、备份、应急和安全措施后，方可由有资格的操作人员执行。

1.1.4 网络管理员的任务

网络管理员负责企业网络的正常运行，保证业务不因网络故障而中断。在网络管理的过程中，要根据管理的网络规模、具体业务要求、组织制度等，结合科学的方法制定一套网络维护方案。在网络维护过程中，主要包括以下任务。

1）应对网络变动

因为人员和业务的调整，网络拓扑结构和功能可能会发生变化，所以网络管理员需要通过修改配置或改变布线来应对网络变动。

2）新设备的安装和配置

此任务包括添加接口、调整链路容量、增加或删除网络设备等。

3）替换故障设备

在设备发生硬件故障后，需要将厂家或备件库提供的新设备进行更换和配置，并进行业务测试。

4）设备配置和软件备份

因为任务更新，所以需要完成设备配置更新，并对软件和配置进行备份。当设备发生严重故障且无法恢复时，可以使用备份系统。

5）排除链路故障

在网络运行过程中，经常会发生链路故障，因此需要网络管理员能够综合分析故障现象，诊断和确定故障原因，解决链路故障。

6）升级软件

在设备需要加载新的软件时，往往需要对原有系统软件进行升级，在确保网络正常运行的情况下，做好升级方案并顺利实施。

7）监控网络

对正在运行的网络设备的各种指示灯的状态、板卡的状态等进行日志收集，通过日志监控网络的运行状态。

8）更新文档

对建设完成的网络进行文档整理，包括 IP 地址、VLAN 规划、路由规划、设备密码等。在每次发生故障或更新设备后，需要对原有文档进行更新，以确保文档的实时性、有效性和完整性。

1.2 网络运行管理制度

为了确保网络系统的稳定和可靠，确保业务系统能够正常运行，各企事业单位均应按照国家相关法律、法规中的要求，建立完善的网络运行管理制度，即通过人、工具、经验等资源，并结合行业特点，形成一套完整的书面制度，以此指导和要求网络管理员及应用人员在网络维护过程中按照规定执行各种操作。

目前，国家制定的相关法规如下。

① 《数据出境安全评估办法》（2022）。

② 《中华人民共和国数据安全法》（2021）、《中华人民共和国个人信息保护法》（2021）。

③ 《互联网域名管理办法》（2017）。

④ 《中华人民共和国网络安全法》（2016）。

⑤ 《计算机信息网络国际联网安全保护管理办法》（2011）、《中华人民共和国计算机信息系统安全保护条例》（2011）。

⑥ 《互联网出版管理暂行规定》（2002）。

⑦ 《计算机病毒防治管理办法》（2000）。

1）网络运行的管理机构

在一般情况下，网络的整体运行和管理都由本单位的网络中心负责。网络中心的主要职责如下。

① 负责网络的正常运行，确保系统的稳定和可靠。

② 负责网络的日常扩容。

③ 负责机房基础设施及服务器、交换机的日常维护巡检，保持机房运行环境的良好状态。

④ 负责网络性能的监控、故障的排除。

⑤ 负责使用人员的技术培训。

⑥ 负责软件的升级。

⑦ 负责设备的维修和调试。

2）网络运行的安全保密制度

① 网络运行管理机构负责网络系统的整体运行，其他未授权用户不能擅自修改网络系统中的任何配置。

② 除网络运行管理机构外，其他用户不能随意监听网络设备的运行情况。

③ 非工作人员不能随意登录和配置网络设备。

④ 定期修改设备密码。

3）其他执法事项

随着互联网的普及，网络安全成为社会关注焦点，网络管理员必须知道相关法律、法规。2023 年 2 月，《工业和信息化部行政执法事项清单（2022 年版）》发布，其中新增了 15 条涉及数据安全的行政执法事项。

具体要求对工业和信息化领域数据处理者开展数据处理活动。

① 落实数据安全保护责任义务及管理措施落实的监督检查。

② 未依照法律、法规的规定，建立健全全流程数据安全管理制度的行政处罚。

③ 未依照法律、法规的规定，组织开展数据安全教育培训的行政处罚。

④ 未依照法律、法规的规定，采取相应的技术措施和其他必要措施，保障数据安全的行政处罚。

⑤ 利用互联网等信息网络开展数据处理活动，未在网络安全等级保护制度的基础上，履行第二十七条数据安全保护义务的行政处罚。

⑥ 未明确数据安全负责人和管理机构，落实数据安全保护责任的行政处罚。

⑦ 未加强风险监测，发现数据安全缺陷、漏洞等风险时，未立即采取补救措施的行政处罚。

⑧ 发生数据安全事件时，未立即采取处置措施的行政处罚。

⑨ 发生数据安全事件时，未按照规定及时告知用户并向有关主管部门报告的行政处罚。

⑩ 未按照规定对其数据处理活动定期开展风险评估，并向有关主管部门报送风险评估报告的行政处罚。

⑪ 报送的风险评估报告未包括处理的重要数据的种类、数量，开展数据处理活动的情况，面临的数据安全风险及其应对措施等的行政处罚。

⑫ 关键信息基础设施的运营者在中华人民共和国境内运营中收集和产生的重要数据的出境安全管理，未落实《中华人民共和国网络安全法》的有关规定的行政处罚。

⑬ 非关键信息基础设施运营者的数据处理者在中华人民共和国境内运营中收集和产生的重要数据的出境安全管理，未落实《数据出境安全评估办法》等有关规定的行政处罚。

⑭ 对数据交易中介服务的机构，未要求数据提供方说明数据来源，审核交易双方的身份，并留存审核、交易记录的行政处罚。

⑮ 未经工业、电信、无线电领域主管机关批准向外国司法或者执法机构提供存储于境内的数据的行政处罚。

素质拓展：新理念 新思路 新办法

党的二十大报告指出："必须坚持问题导向。问题是时代的声音，回答并指导解决问题是理论的根本任务。今天我们所面临问题的复杂程度、解决问题的艰巨程度明显加大，给理论创新提出了全新要求。我们要增强问题意识，聚焦实践遇到的新问题、改革发展稳定存在的深层次问题、人民群众急难愁盼问题、国际变局中的重大问题、党的建设面临的突出问题，不断提出真正解决问题的新理念新思路新办法。"

增值服务

"增值服务"的核心内容是根据客户需要，为客户提供超出常规服务范围，或者采用超出常规服务方法的服务。"增值服务"既包括一般意义上的增值服务，也包括更深层次的延伸服务，如增强客户增值体验的IT服务，能为第三方企业产生区别于其他竞争对手的特色服务。

习题

1. 简述网络管理的主要功能。
2. 如何保证网络的安全管理？
3. 网络管理员的主要任务是什么？
4. 网络运行管理机构的主要职责有哪些？

项目 2

网络故障的排除方法与诊断

内容介绍

网络故障是指在网络系统运行过程中，因为硬件、软件、操作或安全等问题造成网络不能正常运行的情况。造成网络运行故障的原因有很多，如 IP 地址冲突、交换协议故障、路由协议故障、安全攻击、病毒等。因此，掌握网络故障诊断的软/硬件工具的使用方法，学会使用常用的网络测试命令，能够总结一套快速、有效的网络故障排除方法，解决网络故障问题，是每位网络管理员必须具备的能力。

任务安排

任务 1　认识网络故障
任务 2　掌握网络故障的排除方法
任务 3　学会使用网络故障诊断工具

学习目标

◇ 了解网络故障的概念
◇ 熟悉网络结构化故障的排除方法
◇ 了解网络故障诊断工具

素质目标

认识网络技术对数字经济的重要性，认同并维护国家的科教兴国战略；主动负责机房基础设施、服务器、交换机的日常维护和巡检，保持机房运行环境的良好状态；具有社会责任感和社会参与意识，具有 ICT 从业人员高度的使命感及自律性。

2.1 认识网络故障

常见的网络故障如下。

1. 物理故障

1）线路故障

线路故障是指设备之间的连接线缆、线路接口等发生故障，导致网络连接失败。

2）硬件故障

硬件故障是指设备的物理接口出现了问题，如网卡、交换机接口、路由器接口等发生故障。

2. 逻辑故障

1）规划故障

规划故障是指因对协议理解失误，造成规划出错而引起的故障，如 IP 地址冲突、VLAN 协议故障、STP（Spanning Tree Protocol，生成树协议）故障、路由协议故障等。

2）协议故障

协议故障是指在网络通信过程中，因不同厂家产品协议不同而造成的网络连接失败，这需要通过调整协议参数来排除故障。

逻辑故障是最常见的故障之一。除上述两类常见故障外，还有一些是系统负载过高、路由器负载过高等引起的故障。

3. 人为故障

1）操作故障

操作故障是指因用户操作失误造成硬件损坏而引起的故障，如接口被烧、无接地措施、防静电措施不当等。

2）配置故障

配置故障是指在项目实施过程中，因人为配置不当而引起的网络故障，主要表现在不能实现网络所提供的各种服务。

2.2 网络故障的排除方法

网络故障排除方法

网络故障不是一门精密学科，当发生网络故障后，不能保证在一定时间内、任何条件下均能成功诊断并解决所有网络故障问题，有时需要采用多种办法才能解决网络故障问题。目前，常用的网络故障排除方法是结构化故障排除方法。结构化故障排除方法能够系统、全面地分析故障原因，提高故障定位的准确度，减少故障排除的时间，类似于人工智能学习机，当遇到新的故障时能够更快地解决问题。结构化故障排除方法适合零基础网络管理员使用。

2.2.1 结构化故障排除的过程

一般来说，一个网络故障的处理要遵循一个逻辑过程，每步之间都有较强的逻辑关系。图 2-1 所示为结构化故障排除的过程。

图 2-1 结构化故障排除的过程

从图 2-1 中可以看出，结构化故障排除的过程就是一个不断收集信息、分析信息，并进行假设、验证，逐步消除可能的原因，直到最终解决问题的过程。

下面对结构化故障排除过程中涉及的 7 个步骤进行具体介绍。

1）定义故障——描述问题

定义故障是在正确判断故障现象的基础上做出的，因此发现故障的网络管理员要能够清楚地表达故障发生后设备的状态，包括硬件的物理状态、设备板卡的状态、电源的状态、网络互联的状态等。定义故障可以按照以下步骤完成。

① 收集所有故障现象的信息。
② 对现象和问题进行书面详细描述。
③ 尽可能先对重要数据进行备份。
④ 描述网络故障对业务的影响程度。
⑤ 只做记录不下结论。

2）收集信息

在完成定义故障的基础上，确定要收集的信息内容。在收集信息时，需要有充足的工具，如计算机、配置线缆、网络测试仪、万用电表、成品双绞线等。此外，还需要准备好产品书、技术手册、相应的配置软件等。收集信息可以按照以下步骤完成。

① 准备收集信息的工具。
② 收集物理链路的通断情况并记录在表中。
③ 收集发生故障的设备数量和功能并记录在表中。
④ 对原始配置进行备份。
⑤ 收集设备的 IP 地址和管理密码。

⑥ 收集设备的其他参数，如软件版本、设备运行状态、各协议的运行状态等。

3）分析信息

根据收集到的信息，按照网络故障的复杂程度，确定由个人还是由小组来完成任务。虽然发生故障的原因有很多，但是归根到底就是硬件或软件的问题。确切来说，网络故障主要包括网络连接性问题、配置文档问题、网络协议问题、网络超负载问题等。分析过程主要依据故障现象、收集到的信息、网络管理员的技术水平和故障排除经验等做出判断，确定最有可能的假设。分析信息可以按照以下步骤完成。

① 检查并确定所收集信息的完整性和准确性。

② 成立故障排除小组。

③ 确定适合本次网络故障的分析方法。

④ 组织技术专家讨论故障原因。

⑤ 罗列可能的故障原因。

4）消除可能性小的原因

根据罗列的可能的故障原因，逐条进行排查，组织研讨会进行讨论，消除可能性小的原因，尽量缩小引起网络故障的原因范围。消除可能性小的原因可以按照以下步骤完成。

① 根据可能性对网络故障原因进行排序。

② 组织研讨会，逐个讨论原因。

③ 缩小可能引起故障的原因范围。

④ 确定最有可能的故障原因。

5）推断根本性故障原因

在确定最有可能的故障原因后，根据实际情况，先提出最有可能的原因假设，再根据这个假设进行相关准备工作，包括设备备件、线路、工具、软件等方面的准备，并给出故障解决方案，为解决问题做好准备。推断根本性故障原因可以按照以下步骤完成。

① 确定最有可能的故障原因。

② 做好准备工作。

③ 讨论方案。

④ 协调业务要求，选择故障排除的最佳时间。

6）验证推断——测试假设

这个阶段是为最终解决问题做最后的工作。如果有测试环境，那么可以在测试环境中进行测试，这样可以降低因为测试失败造成更大网络故障的可能性。如果没有测试环境，必须在真实环境中完成测试，那么需要做好以下工作。

① 审查测试方案。

② 确定测试人员。

③ 准备测试工具。

④ 确定测试时间。

⑤ 确定测试失败的回退方案。

⑥ 对业务进行测试。
⑦ 形成正确测试结果。

7）解决问题

作为网络故障排除的最后一个阶段，解决问题主要完成故障的处理，把网络系统恢复到正常运行状态。解决问题主要完成以下任务。

① 依据测试结果，修订出最终的故障排除方案。
② 确定人员和时间。
③ 准备解决问题的各项条件。
④ 组织并实施故障排除。
⑤ 实施完成后进行业务测试。
⑥ 更新维护文档。
⑦ 总结故障案例。

2.2.2 故障排除的几种具体方法

结构化故障排除方法可以作为一个准则来解决网络故障，但是在实际解决网络故障的过程中还需要采取一些具体的方法。

1）自上而下

自上而下的解决方法是把网络系统看作一个 TCP/IP 模型，从应用层到物理层逐层进行假设、验证。这种方法一般比较适合解决应用类故障。

2）自下而上

自下而上的解决方法是从 TCP/IP 模型的物理层出发，逐步扩展到应用层，在这个过程中检查相关网络元素能否正常运行。通常采用测试命令来完成网络层以下的功能确认。这种方法适用于规模较小，并且之前发生的大多数网络故障都和物理层有关的网络。如果是在大型网络中，那么它将是一个耗时的过程，因为需要大量的时间来确认网络的有效范围并逐步采取措施。

3）分而治之

分而治之又被称为中间法，一般是从网络层开始排除，可以执行如 ping 的测试命令，根据测试的结果来判断应该往上还是往下，能够迅速地确定故障排除的大方向。因此，分而治之的解决方法被认为是一个非常有效的故障排除方法。

4）路径排查

路径排查的解决方法是最基本的故障排除方法之一，是对其他排除方法的补充。路径排查方法的原理是首先确定数据包的实际路径，以及从源地址到目的地址之间的链路和设备，然后根据可达路径将故障的范围确定在某个链路或设备上，从而加快排查速度。

2.3 网络故障诊断工具

2.3.1 故障诊断工具

在故障排除过程中，选择合适的故障诊断工具能够起到事半功倍的效果。故障诊断工具主要包括硬件工具、软件工具和各种测试命令。

1. 故障诊断的硬件工具

硬件工具在处理物理故障时非常有效，特别是在测试链路和连接电缆的连通性时必须使用硬件工具来完成。硬件工具主要包括以下设备。

1）网络测试仪

网络测试仪通常又被称为专业网络测试仪或网络检测仪，是一种便携、可视的智能检测设备，用于检测网络的物理层、数据链路层、网络层运行状况，主要适用于局域网故障检测维护和综合布线施工，如图2-2所示。

图2-2 网络测试仪

网络测试仪的使用可以极大地减少网络管理员排查网络故障的时间，从而提高综合布线施工人员的工作效率，加速工程进度，提高工程质量。该类设备在国外已经被广泛应用，是网络检测和网络施工过程中必不可少的工具。相对于国外，该类设备在国内的使用范围仍然有限，一般以普通测试仪为主。这主要是因为国内用户对该类产品的认识度还不够，对网络故障的敏感度不高。国外的网络测试仪厂商有福禄克、安捷伦和理想等，国内的网络测试仪厂商有德利、中创信测等。

网络测试仪按网络传输介质不同可以分为无线网络测试仪和有线网络测试仪两类。

（1）无线网络测试仪。

无线网络测试仪主要针对无线路由和无线AP（无线访问节点）进行检测，可以检测出无线网络中连接的终端和无线信号的强度，能有效地管理网络中的节点，增强网络安全。例如，福禄克的NETSCOUT Aircheck可以检测信号的强度、信噪比、噪声等信号问题，可以检测信道接入点的数量和信道利用率等信道问题，可以检测SSID下有多少AP等终端数量问题，还可以检测无线、有线连通性，信道吞吐和定位AP等问题，如图2-3所示。当然，该类仪器价格相对较高。

图 2-3　无线网络测试仪

（2）有线网络测试仪。

有线网络中常见的传输介质包括双绞线、光纤和同轴电缆。市场上普通的有线网络测试仪通常是指双绞线网络测试仪。高档的有线网络测试仪通常通过多连接适配器接口解决测试双绞线、光纤等不同网络介质的网络状况问题。双绞线网络测试仪主要有以下功能。

① 链路识别：可判断网络链路速度，包括 10Mbps、100Mbps、1Gbps 的网络，同时可判断网络的半双工/全双工状态。

② ping：可进行 IP 连通性测试。

③ POE 检测：检测电压的可用性、电压水平和以太网供电线是否符合 IEEE 802.3af 规范。

④ 识别接口：识别互连的链路连接在网络设备的哪个接口上。

⑤ 电缆诊断：测试布线图和电缆长度；检测错线、短路、串绕线对或开路问题。

⑥ 定位电缆：利用 IntelliTone 数字音频技术、集线器闪烁功能或可选的电缆 ID 来定位配线板或墙壁插座中的电缆。

⑦ 数据管理：将链路、ping、POE、接口和电缆的测试结果保存到设备中，并通过 USB 电缆上传到计算机中。

2）万用电表

万用电表简称万用表，是一种多功能、多量程、便于携带的电子仪表。它可以用来测量直流电流/电压、交流电流/电压、电阻、音频电平和晶体管直流放大倍数等物理量。万用表由表头、测量线路、转换开关和测试表笔等组成。其中，万用表各组件的功能如下。

① 刻度尺：显示各种被测量的数值及范围。

② 量程选择开关：根据具体情况转换不同的量程、不同的物理量。

③ 机械零位调节旋钮：用来校准指针的机械零位。

④ 欧姆挡零位调节旋钮：用来进行电气零位调节。

⑤ 插孔或接线柱：用来连接测试表笔。

3）其他硬件

（1）交叉电缆。

交叉电缆可以绕过网络直接对计算机的通信能力进行隔离和测试。

（2）示波器。

示波器是一种以时间为单位测量信号电压值，并在显示器上显示结果的电子装置。

2. 故障诊断的软件工具

随着网络应用的发展，有很多软件工具可以帮助网络管理员进行网络故障排除。软件工具主要包括以下几种。

1）网络监视器

网络监视器是操作系统（如 Microsoft Windows Server 系列）自带的实用程序。网络管理员可以通过网络监视器对网络中一个或多个网络监视终端进行行为控制，还可以使用网络监视器捕获和查看网络的通信模式和问题。网络监视器通常包括以下两个工作步骤。

第一步：启动网络监视器，此时将显示捕获窗口。该窗口包含 4 个帧，即关系图、会话统计信息、站统计信息、总体统计信息。

第二步：在捕获信息后，可以立即通过网络监视器中的帧查看器窗口进行查看，也可以将其保存到一个文件中进行分析。

2）协议分析器

网络协议分析是指通过程序分析网络数据包的协议头和尾，从而了解数据包在产生和传输过程中的行为及其他相关信息。协议分析器就是包含该程序的软件和设备。协议分析器是一种用于监督和跟踪网络活动的诊断工具。网络运行和维护的很多方面都可以使用协议分析器，如监视网络流量、分析数据包、监视网络资源的利用、执行网络安全操作规则、分析鉴定网络数据，以及诊断并修复网络问题等。

协议分析器可以是计算机上运行的软件，也可以是包含特殊线路板和软件的便携设备。协议分析器主要有以下功能。

① 显示网络上传输信息分组的类型信息。网络管理员可以通过监督这些分组来监督网络的安全性，确定失效情况，或者监督和优化一个网络。

② 在一个互联网络上查询所有节点，或者在任何一个特定节点与所有其他节点之间进行点对点通信检测，确定整个互联网络的配置。

③ 从一个或所有节点分析关键数据，或者根据预先定义的一组阈值报告不正常的活动。

④ 显示性能数据，如通信量和被服务的分组。

⑤ 提供关于网络有效性、网络性能、可能的硬件错误、噪声问题和应用软件问题的一些有用信息。

2.3.2 常用的网络测试命令

1. ping 命令

ping 命令是操作系统自带的测试命令，在 Windows、UNIX 和 Linux 系统中都有这个命令，用来检查网络是否通畅。ping 命令的基本行为是发送一个 ICMP（Internet Control Message Protocol，互联网控制报文协议）请求消息给目的地，并报告是否收到所希望的 ICMP 回显应答。作为一名网络管理员，ping 命令是第一个必须掌握的 DOS 命令。ping 命令的原理是先利用网络计算机 IP 地址的唯一性，给目标 IP 地址发送一个数据包，再要求对方返回一个同样大小的数据包来确定两台计算机是否连通、时延是多少。

ping 命令的格式如下：

```
ping [-t] [-a] [-n count] [-l size] [-f] [-i TTL] [-v TOS][-r count] [-s count] [[-j host-list] | [-k host-list]][-w timeout] [-R] [-S srcaddr] [-c compartment] [-p][-4] [-6] target_name
```

读者可以运行"ping /?"命令来调用帮助文档，以了解各参数格式及其功能。

ping 命令常见参数的说明如下。

-t：ping 指定的主机，直到人为停止。若要查看统计信息并继续操作，则按 Ctrl+Break 组合键；若要停止，则按 Ctrl+C 组合键。

-a：将地址解析为主机名。

-n count：要发送的回显请求数量。可以自己定义发送的数量，默认值为 4。

-l size：发送缓冲区大小。

-f：在数据包中设置"不分段"标记（仅适用于 IPv4）。

-i TTL：生存时间。

-v TOS：服务类型（仅适用于 IPv4。该参数已被弃用，对 IP 标头中的服务类型字段没有任何影响）。

-r count：记录跃点计数的路由（仅适用于 IPv4）。

-s count：跃点计数的时间戳（仅适用于 IPv4）。

-j host-list：与主机列表一起使用的松散源路由（仅适用于 IPv4）。

-k host-list：与主机列表一起使用的严格源路由（仅适用于 IPv4）。

-w timeout：等待每次回复的超时时间（毫秒）。

-R：同样使用路由标头测试反向路由（仅适用于 IPv6）。若使用此标头，则某些系统可能丢弃回显请求。

-S srcaddr：要使用的源地址。

-c compartment：路由隔离舱标识符。

-p：为 ping Hyper-V 网络虚拟化提供程序地址。

-4：强制使用 IPv4。

-6：强制使用 IPv6。

ping 命令除了上述参数，还有其他扩展命令，如 pathping 或 ping + tracert 组合。读者可以自行体验 ping 命令的丰富功能。

2. ipconfig 命令

ipconfig 命令用于在 Windows 环境中检测 TCP/IP 的设置是否正确。

ipconfig 命令的格式如下：

```
ipconfig [/allcompartments] [/? | /all |/renew [adapter] | /release [adapter]
|/renew6 [adapter] | /release6 [adapter] |/flushdns | /displaydns | /registerdns |
/showclassid adapter |/setclassid adapter [classid] |/showclassid6 adapter
|/setclassid6 adapter [classid] ]
```

读者可以运行"ipconfig /? "命令来调用帮助文档，以了解各参数格式及其功能。

ipconfig 命令常见参数的说明如下。

ipconfig /all：显示本机 TCP/IP 配置的详细信息。

ipconfig /renew：DHCP 客户端手动向服务器刷新请求。

ipconfig /release：DHCP 客户端手动释放 IP 地址。

ipconfig /flushdns：清除本地 DNS 缓存内容。

ipconfig /displaydns：显示本地 DNS 内容。

ipconfig /registerdns：DNS 客户端手动向服务器注册。

3. tracert 命令

tracert 是路由跟踪命令，用于确定数据包访问目标所采取的路径。tracert 命令通过 IP 生存时间（TTL）字段和 ICMP 错误消息来确定从一个主机到网络上其他主机的路由。数据包在经过每台路由器转发时 TTL 递减 1，所以 TTL 是有效的跃点计数。当数据包上的 TTL 到达 0 时，路由器应该将"ICMP 已超时"的消息发送回原主机。tracert 命令先发送 TTL 为 1 的回显数据包，并在随后的每次发送过程中都将 TTL 递增 1，直到目标响应或 TTL 达到最大值，从而确定路由。

tracert 命令的格式如下：

```
tracert [-d] [-h maximum_hops] [-j computer-list] [-w timeout] target_name
```

读者可以运行"tracert /? "命令来调用帮助文档，以了解各参数格式及其功能。

tracert 命令常见参数的说明如下。

-d：指定不将地址解析成主机名。

-h maximum_hops：指定搜索目标的最大跃点数。

-j computer-list：指定 tracert 实用程序数据包所采用路径中的路由器接口列表。

-w timeout：每次应答等待 timeout 指定的毫秒数。

target_name：目标主机的名称。

4. netstat 命令

netstat 命令是控制台命令，是一个非常有用的监控 TCP/IP 网络的工具。它可以显示路由表、实际的网络连接和每个网络接口设备的状态信息。netstat 命令用于显示与 IP、TCP、UDP 和 ICMP 相关的统计数据，一般用于检验本机各接口的网络连接情况。

netstat 命令的格式如下：

```
netstat [-a] [-b] [-e] [-f] [-n] [-o] [-p proto] [-r] [-s] [-t] [interval]
```

读者可以运行"netstat /?"命令来调用帮助文档,以了解各参数格式及其功能。

netstat 命令常见参数的说明如下。

-a:显示所有连接和侦听接口。

-b:显示在创建每个连接或侦听接口时涉及的可执行程序。

-e:显示以太网统计。此参数可以与-s 参数结合使用。

-f:显示外部地址的完全限定域名(FQDN)。

-n:以数字形式显示地址和接口号。

-o:显示拥有的与每个连接关联的进程 ID。

-p proto:显示 proto 指定的协议的连接。proto 可以是 TCP、TCPv6、UDP 或 UDPv6 中的任何一个。如果结合使用-p 与-s 参数来显示每个协议的统计,那么 proto 可以是 IP、IPv6、ICMP、ICMPv6、TCP、TCPv6、UDP 或 UDPv6 中的任何一个。

-r:显示路由表。

-s:显示每个协议的统计。在默认情况下,使用-s 参数可以显示 IP、IPv6、ICMP、ICMPv6、TCP、TCPv6、UDP 和 UDPv6 的统计。如果结合使用-s 与-p 参数,那么-p 参数可用于指定默认的子网。

-t:显示当前连接/卸载状态。

interval:重新显示选定的统计,每次显示之间间隔的秒数用 interval 来指定。按 Ctrl+C 组合键可以停止重新显示统计。

素质拓展:新网络 新科技 新赋能

在"智联世界 元生无界"2022 世界人工智能大会(WAIC)上,中国联通智慧医疗团队独立研发的基于模型的条件独立性检验方法荣获了大赛一等奖。中国联通是一家"管道运营商",怎么变为科技创新公司了?

2022 年,公司董事长刘烈宏在一次季度业绩说明会上介绍了中国联通近年来快速推进 900MHz 低频网打底网建设,提升了农村及边远地区的网络服务。

一是携手中国电信建成全球最大的共建共享 5G 网络,中国联通 900MHz 低频网在 2022 年年底实现农村覆盖 17 万个站点,将快速提升农村 5G 覆盖水平,助力数字乡村建设。

二是千兆宽带精品网覆盖超过 4.5 亿个家庭,联通智家惠及千家万户。

三是升级打造低时延、高可靠、智能化的政企精品网,覆盖超过 300 个城市。

四是依托全光底座,引领 IPv6+创新,构建"5+4+31+X"多级算力布局,打造云网边协同的算力网。

2022 年以来,中国联通相继完成了北京冬奥会和冬残奥会等级的通信保障任务。

同时,中国联通携手雅戈尔共同打造 5G 全连接工厂,实现宁波总部、吉林珲春、云南瑞丽三地的云上管理、一体联动,一件西装的高级定制时间从原来的 15 天缩短到 5 天,批量订单生产周期缩短了 35%,单工位生产效率提高了 25%。

"新网络、新科技、新赋能"是中国联通公司的成功之道。

增值服务

通过网络故障的排除，售后工程人员基本熟悉了企业业务网络。在排除故障的过程中，售后工程人员如果发现非故障表现的潜在问题或隐患，应提出合理化建议或措施，让用户感受到在本次服务之外的技术服务的温暖，从而增强被服务的体验感。

例如，在移交故障修复文档时，若售后工程人员关注到本项目设备没有开启远程维护功能，则要提醒企业业务主管开启远程维护功能。这样便于厂家售后工程人员的远程维护，提高工作效率，降低企业运维成本。

习题

1. 简述常见的网络故障。
2. 简述结构化故障排除的过程。
3. 请尝试使用 ping 命令测试你的主机与百度官网能否正常通信。
4. 请尝试使用 "tracert www.baidu.com" 命令测试你的主机与百度官网之间的路由信息。

项目 3

局域网 VLAN 的故障排除

内容介绍

某公司已经建成内部局域网，所有交换机已完成 VLAN 配置。最近因为新增业务，所以需要在交换机上完成新业务的接入，即在交换机 S2、S3 上增加新的 VLAN，并确保业务的互联互通。

任务安排

任务 1　针对新业务更新网络配置
任务 2　进行网络更新过程中的故障分析与排除

学习目标

- ◇ 了解 VLAN 常见故障的原因
- ◇ 掌握故障排除的思路
- ◇ 学会结构化故障排除方法
- ◇ 学会 VLAN 相关故障排除及文档更新的方法

素质目标

具有一定的资料收集与分析能力，良好的沟通协调能力，以及较强的独立分析问题和解决问题能力，并具有实事求是、与时俱进的科学态度。

3.1 VLAN 配置分析与实施

发现故障

网络管理员李工接到更新公司网络的任务，需要对交换机配置做新的调整方案，因此可执行以下操作。

① 查看原有交换机的配置，落实已有的 VLAN 和 IP 地址信息，查看新增终端所经过的交换机，并进行标识。

② 规划新的 VLAN 和 IP 地址信息，规划新增终端在交换机上的互联接口。

③ 对交换机进行配置，包括终端接入交换机和跨 VLAN 的中间交换机。

④ 进行新增终端的互联互通测试，并分析测试过程中数据包的转发过程。

公司原有网络拓扑结构如图 3-1 所示。

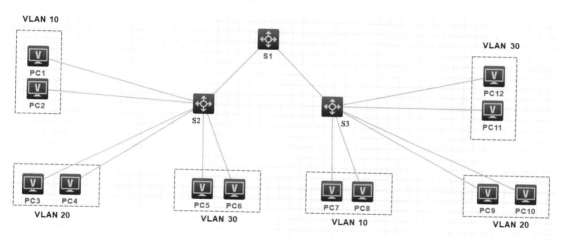

图 3-1 公司原有网络拓扑结构

1）查看原有交换机的配置

各交换机的 VLAN 配置及终端 PC 的 IP 地址规划如表 3-1[①]所示。

表 3-1 各交换机的 VLAN 配置及终端 PC 的 IP 地址规划

交换机	接口类型	业务名称	接口	对端	IP 地址	子网掩码
S1	VLAN 10	yewu1			192.168.10.254	255.255.255.0
	VLAN 20	yewu2			192.168.20.254	255.255.255.0
	VLAN 30	yewu3			192.168.30.254	255.255.255.0
	trunk		F0/24	S2		
	trunk		F0/23	S3		

---

① 本书使用 F 表示 FortyGigE。

续表

交换机	接口类型	业务名称	接口	对端	IP 地址	子网掩码
S2	VLAN 10	yewu1	F0/1	PC1	192.168.10.1	255.255.255.0
			F0/2	PC2	192.168.10.2	255.255.255.0
	VLAN 20	yewu2	F0/6	PC3	192.168.20.1	255.255.255.0
			F0/7	PC4	192.168.20.2	255.255.255.0
	VLAN 30	yewu3	F0/11	PC5	192.168.30.1	255.255.255.0
			F0/12	PC6	192.168.30.2	255.255.255.0
	trunk		F0/24	S1		
S3	VLAN 10	yewu1	F0/1	PC7	192.168.10.6	255.255.255.0
			F0/2	PC8	192.168.10.7	255.255.255.0
	VLAN 20	yewu2	F0/6	PC9	192.168.20.6	255.255.255.0
			F0/7	PC10	192.168.20.7	255.255.255.0
	VLAN 30	yewu3	F0/11	PC11	192.168.30.6	255.255.255.0
			F0/12	PC12	192.168.30.7	255.255.255.0
	trunk		F0/23	S1		

2）规划并配置新的 VLAN 和 IP 地址

如图 3-1 所示，公司原有网络由 3 台交换机（S1、S2、S3）组成，在交换机 S2、S3 上分别有 yewu1、yewu2 和 yewu3 共 3 个 VLAN（VLAN 10、VLAN 20、VLAN 30），S1 连接 S2 和 S3，连接接口为 trunk 接口，S2 和 S3 上相同 VLAN 间的通信跨 S1 完成。

由于业务扩展，在原有网络拓扑结构上完成新增 VLAN 的 IP 地址的配置后，形成的新网络拓扑结构如图 3-2 所示。在新网络拓扑结构中，VLAN 40 和 VLAN 50 为新增的 VLAN 和终端。其中，VLAN 40（包括 S2:F0/16～F0/19 和 S3:F0/16～F0/19，其中 S2:F0/16 对应 PC_13，S2:F0/17 对应 PC_14，S3:F0/16 对应 PC_17，S3:F0/17 对应 PC_18）和 VLAN 50（包括 S2:F0/20～F0/23 和 S3:F0/20～F0/23，其中 S2:F0/20 对应 PC_15，S2:F0/21 对应 PC_16，S3:F0/20 对应 PC_19，S3:F0/21 对应 PC_20）为新增的 VLAN，PC_13～PC_20 为新增的终端，要求 S2 和 S3 上新增的 VLAN 能够通过 S1 实现互联互通。

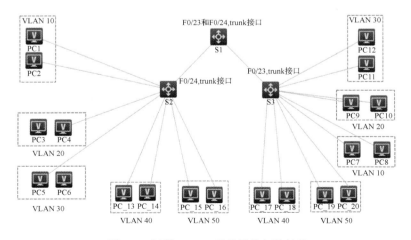

图 3-2 新增 VLAN 后的网络拓扑结构

3）测试

在增加配置后,进行 PC 之间的连通性测试,发现存在如表 3-2 所示的故障现象。

表 3-2 连通性测试结果

测试序号	交换机 S2	交换机 S3	测试方法	预期测试结果	实际测试结果	是否发生故障
1	PC_13、PC_14		ping 命令	成功	失败	是
2	PC_15	PC_20	ping 命令	成功	失败	是
3	PC_16	PC_20	ping 命令	成功	失败	是

由表 3-2 可知,共有 3 项测试失败,因此表明本次操作没有成功完成项目目标。是什么原因造成的故障现象呢?是规划设计的问题,还是操作的问题,还是 VLAN 概念没有理解清楚的问题?下面进行故障原因的分析。

3.2 VLAN 故障分析与排除

排除故障

1. 故障分析方法

根据结构化故障排除思路,严格执行故障排除的操作步骤。首先,确定故障现象并进行详细记录;其次,收集设备信息,本项目主要收集交换机的配置信息、IP 地址、接口信息、VLAN 信息等;再次,在收集完信息后,结合 VLAN 的实现原理进行综合分析,确定并罗列可能存在的故障点;最后,针对故障点分析出最有可能的故障原因,并对这个原因进行故障排除。

2. 分析故障点

从 VLAN 的实现及技术原理上分析,可能存在以下故障点。

① 缺少 VLAN 或存在错误的 VLAN。
② VLAN 中继协议(VTP)配置错误。
③ 在错误的 VLAN 上设置了接入接口。
④ 缺少树干或存在配置错误的树干。
⑤ 与本地 VLAN 不匹配。

由于本次配置通过模拟环境来实现,因此可以忽略物理问题和设备问题(若在实际环境下,则应该逐步排查)。这里主要从 VLAN 的技术原理及 VLAN 的配置和操作方面排除故障。

1）查看所有交换机的配置

(1)任务目标。

① 查看原有交换机的配置,确定已有的配置信息。
② 查看新增交换机的配置,确定已有的配置信息。

(2)任务所需设备。

① 一台装有超级终端软件或 Telnet 软件的计算机,同时确定访问所需的用户名和口令。
② 配置线缆。

③ 笔和纸，用于记录相关信息。

（3）具体实施。

① 使用超级终端软件或 Telnet 软件连接交换机，并使用 disp vlan all 命令查看原有交换机 S2 和 S3 的 VLAN 配置信息和空闲接口信息。

使用 disp vlan all 命令查看交换机 S2 的 VLAN 配置信息（通常，通过查看工程竣工文档来获得这些信息）：

```
[S2]disp vlan all
VLAN ID: 1
VLAN type: Static
Route interface: Not configured
Description: VLAN 0001
Name: VLAN 0001
Tagged ports:    None
Untagged ports:
    FortyGigE1/0/53           FortyGigE1/0/54
    GigabitEthernet1/0/16     GigabitEthernet1/0/17
    GigabitEthernet1/0/18     GigabitEthernet1/0/19
    GigabitEthernet1/0/20     GigabitEthernet1/0/21
    GigabitEthernet1/0/22     GigabitEthernet1/0/23
    GigabitEthernet1/0/24     GigabitEthernet1/0/25
    GigabitEthernet1/0/26     GigabitEthernet1/0/27
    GigabitEthernet1/0/28     GigabitEthernet1/0/29
    GigabitEthernet1/0/30     GigabitEthernet1/0/31
    GigabitEthernet1/0/32     GigabitEthernet1/0/33
    GigabitEthernet1/0/34     GigabitEthernet1/0/35
    GigabitEthernet1/0/36     GigabitEthernet1/0/37
    GigabitEthernet1/0/38     GigabitEthernet1/0/39
    GigabitEthernet1/0/40     GigabitEthernet1/0/41
    GigabitEthernet1/0/42     GigabitEthernet1/0/43
    GigabitEthernet1/0/44     GigabitEthernet1/0/45
    GigabitEthernet1/0/46     GigabitEthernet1/0/47
    GigabitEthernet1/0/48
    Ten-GigabitEthernet1/0/49
    Ten-GigabitEthernet1/0/50
    Ten-GigabitEthernet1/0/51
    Ten-GigabitEthernet1/0/52……
VLAN ID: 10
VLAN type: Static
Route interface: Not configured
Description: VLAN 0010
Name: yewu1
Tagged ports:
    GigabitEthernet1/0/23     GigabitEthernet1/0/24
Untagged ports:
```

```
    GigabitEthernet1/0/1           GigabitEthernet1/0/2
    GigabitEthernet1/0/3           GigabitEthernet1/0/4
    GigabitEthernet1/0/5

VLAN ID: 20
VLAN type: Static
Route interface: Not configured
Description: VLAN 0020
Name: yewu2
Tagged ports:
    GigabitEthernet1/0/23          GigabitEthernet1/0/24
Untagged ports:
    GigabitEthernet1/0/6           GigabitEthernet1/0/7
    GigabitEthernet1/0/8           GigabitEthernet1/0/9
    GigabitEthernet1/0/10

VLAN ID: 30
VLAN type: Static
Route interface: Not configured
Description: VLAN 0030
Name: yewu3
Tagged ports:
    GigabitEthernet1/0/23          GigabitEthernet1/0/24
Untagged ports:
    GigabitEthernet1/0/11          GigabitEthernet1/0/12
    GigabitEthernet1/0/13          GigabitEthernet1/0/14
    GigabitEthernet1/0/15
```

使用 disp vlan all 命令查看交换机 S3 的 VLAN 配置信息：

```
[S3]disp vlan all
VLAN ID: 1
VLAN type: Static
Route interface: Not configured
Description: VLAN 0001
Name: VLAN 0001
Tagged ports:   None
Untagged ports:
    FortyGigE1/0/53                FortyGigE1/0/54
    GigabitEthernet1/0/16          GigabitEthernet1/0/17
    GigabitEthernet1/0/18          GigabitEthernet1/0/19
    GigabitEthernet1/0/20          GigabitEthernet1/0/21
    GigabitEthernet1/0/22          GigabitEthernet1/0/23
    GigabitEthernet1/0/24          GigabitEthernet1/0/25
    GigabitEthernet1/0/26          GigabitEthernet1/0/27
    GigabitEthernet1/0/28          GigabitEthernet1/0/29
    GigabitEthernet1/0/30          GigabitEthernet1/0/31
```

```
    GigabitEthernet1/0/32         GigabitEthernet1/0/33
    GigabitEthernet1/0/34         GigabitEthernet1/0/35
    GigabitEthernet1/0/36         GigabitEthernet1/0/37
    GigabitEthernet1/0/38         GigabitEthernet1/0/39
    GigabitEthernet1/0/40         GigabitEthernet1/0/41
    GigabitEthernet1/0/42         GigabitEthernet1/0/43
    GigabitEthernet1/0/44         GigabitEthernet1/0/45
    GigabitEthernet1/0/46         GigabitEthernet1/0/47
    GigabitEthernet1/0/48
    Ten-GigabitEthernet1/0/49
    Ten-GigabitEthernet1/0/50
    Ten-GigabitEthernet1/0/51
    Ten-GigabitEthernet1/0/52

VLAN ID: 10
VLAN type: Static
Route interface: Not configured
Description: VLAN 0010
Name: yewu1
Tagged ports:
    GigabitEthernet1/0/23         GigabitEthernet1/0/24
Untagged ports:
    GigabitEthernet1/0/1          GigabitEthernet1/0/2
    GigabitEthernet1/0/3          GigabitEthernet1/0/4
    GigabitEthernet1/0/5

VLAN ID: 20
VLAN type: Static
Route interface: Not configured
Description: VLAN 0020
Name: yewu2
Tagged ports:
    GigabitEthernet1/0/23         GigabitEthernet1/0/24
Untagged ports:
    GigabitEthernet1/0/6          GigabitEthernet1/0/7
    GigabitEthernet1/0/8          GigabitEthernet1/0/9
    GigabitEthernet1/0/10

VLAN ID: 30
VLAN type: Static
Route interface: Not configured
Description: VLAN 0030
Name: yewu3
Tagged ports:
    GigabitEthernet1/0/23         GigabitEthernet1/0/24
```

```
Untagged ports:
   GigabitEthernet1/0/11        GigabitEthernet1/0/12
   GigabitEthernet1/0/13        GigabitEthernet1/0/14
   GigabitEthernet1/0/15
```

② 根据以上配置信息，整理出交换机 S2 和 S3 的 VLAN 配置信息和空闲接口信息，并将这些信息填入表 3-3[①]。

表 3-3　交换机 S2 和 S3 的 VLAN 配置信息和空闲接口信息

交换机	VLAN ID	空闲接口
S2	1、10、20、30	F1/0/53、F1/0/54；G1/0/16～G1/0/52
S3	1、10、20、30	F1/0/53、F1/0/54；G1/0/16～G1/0/52

③ 使用软件查看终端 IP 地址配置，如表 3-4 所示。

表 3-4　终端 IP 地址配置

交换机	接口类型	业务名称	接口	对端	IP 地址	子网掩码
S1	VLAN 10	yewu1			192.168.10.254	255.255.255.0
	VLAN 20	yewu2			192.168.20.254	255.255.255.0
	VLAN 30	yewu3			192.168.30.254	255.255.255.0
	trunk		G1/0/23	S3		
	trunk		G1/0/24	S2		
S2	VLAN 10	yewu1	G1/0/1	PC1	192.168.10.1	255.255.255.0
			G1/0/2	PC2	192.168.10.2	255.255.255.0
			G1/0/3			
			G1/0/4			
			G1/0/5			
	VLAN 20	yewu2	G1/0/6	PC3	192.168.20.1	255.255.255.0
			G1/0/7	PC4	192.168.20.2	255.255.255.0
			G1/0/8			
			G1/0/9			
			G1/0/10			
	VLAN 30	yewu3	G1/0/11	PC5	192.168.30.1	255.255.255.0
			G1/0/12	PC6	192.168.30.2	255.255.255.0
			G1/0/13			
			G1/0/14			
			G1/0/15			
	trunk		G1/0/24			

---

① 本书使用 G 表示 GigabitEthernet。

续表

交换机	接口类型	业务名称	接口	对端	IP 地址	子网掩码
S3	VLAN 10	yewu1	G1/0/1	PC7	192.168.10.6	255.255.255.0
			G1/0/2	PC8	192.168.10.7	255.255.255.0
			G1/0/3			
			G1/0/4			
			G1/0/5			
	VLAN 20	yewu2	G1/0/6	PC9	192.168.20.6	255.255.255.0
			G1/0/7	PC10	192.168.20.7	255.255.255.0
			G1/0/8			
			G1/0/9			
			G1/0/10			
	VLAN 30	yewu3	G1/0/11	PC11	192.168.30.6	255.255.255.0
			G1/0/12	PC12	192.168.30.7	255.255.255.0
			G1/0/13			
			G1/0/14			
			G1/0/15			
	trunk		G1/0/24			

新增的 VLAN 40 和 VLAN 50 分别使用 192.168.40.0/24 和 192.168.50.0/24 网段，各终端的 IP 地址没有错误。目前交换机 S2 和 S3 上已有 VLAN 10、VLAN 20 和 VLAN 30，并且空闲接口为 F1/0/53、F1/0/54 及 G1/0/16～G1/0/52，现有 IP 地址配置合理。

④ 查看新增 VLAN 后交换机的配置信息和 VLAN 配置信息。

查看新增 VLAN 后交换机 S2 和 S3 的 VLAN 配置信息和空闲接口信息。

使用 disp vlan all 命令查看交换机 S2 的 VLAN 配置信息：

```
[S2]disp vlan all
VLAN ID: 1
VLAN type: Static
Route interface: Not configured
Description: VLAN 0001
Name: VLAN 0001
Tagged ports:   None
Untagged ports:
  FortyGigE1/0/53              FortyGigE1/0/54
  GigabitEthernet1/0/23        GigabitEthernet1/0/24
  GigabitEthernet1/0/25        GigabitEthernet1/0/26
  GigabitEthernet1/0/27        GigabitEthernet1/0/28
  GigabitEthernet1/0/29        GigabitEthernet1/0/30
  GigabitEthernet1/0/31        GigabitEthernet1/0/32
  GigabitEthernet1/0/33        GigabitEthernet1/0/34
  GigabitEthernet1/0/35        GigabitEthernet1/0/36
  GigabitEthernet1/0/37        GigabitEthernet1/0/38
  GigabitEthernet1/0/39        GigabitEthernet1/0/40
  GigabitEthernet1/0/41        GigabitEthernet1/0/42
```

```
        GigabitEthernet1/0/43          GigabitEthernet1/0/44
        GigabitEthernet1/0/45          GigabitEthernet1/0/46
        GigabitEthernet1/0/47          GigabitEthernet1/0/48
        Ten-GigabitEthernet1/0/49
        Ten-GigabitEthernet1/0/50
        Ten-GigabitEthernet1/0/51
        Ten-GigabitEthernet1/0/52

 VLAN ID: 10
 VLAN type: Static
 Route interface: Not configured
 Description: VLAN 0010
 Name: yewu1
 Tagged ports:
     GigabitEthernet1/0/23          GigabitEthernet1/0/24
 Untagged ports:
     GigabitEthernet1/0/1           GigabitEthernet1/0/2
     GigabitEthernet1/0/3           GigabitEthernet1/0/4
     GigabitEthernet1/0/5

 VLAN ID: 20
 VLAN type: Static
 Route interface: Not configured
 Description: VLAN 0020
 Name: yewu2
 Tagged ports:
     GigabitEthernet1/0/23          GigabitEthernet1/0/24
 Untagged ports:
     GigabitEthernet1/0/6           GigabitEthernet1/0/8
     GigabitEthernet1/0/9           GigabitEthernet1/0/10
     GigabitEthernet1/0/12          //缺少G1/0/7接口

 VLAN ID: 30
 VLAN type: Static
 Route interface: Not configured
 Description: VLAN 0030
 Name: yewu3
 Tagged ports:
     GigabitEthernet1/0/23          GigabitEthernet1/0/24
 Untagged ports:
     GigabitEthernet1/0/7           GigabitEthernet1/0/11
     GigabitEthernet1/0/13          GigabitEthernet1/0/14
     GigabitEthernet1/0/15          //多了G1/0/7接口
```

```
VLAN ID: 40
VLAN type: Static
Route interface: Not configured
Description: VLAN 0040
Name: yewu4
Tagged ports:    None                   //trunk 接口未允许 VLAN 通过
Untagged ports:
   GigabitEthernet1/0/16     GigabitEthernet1/0/18
   GigabitEthernet1/0/19     //缺少 G1/0/17 接口

VLAN ID: 50
VLAN type: Static
Route interface: Not configured
Description: VLAN 0050
Name: yewu5
Tagged ports:    None                   //trunk 接口未允许 VLAN 通过
Untagged ports:
   GigabitEthernet1/0/17     GigabitEthernet1/0/20
   GigabitEthernet1/0/21     GigabitEthernet1/0/22    //多了 G1/0/17 接口
```

使用 disp vlan all 命令查看交换机 S3 的 VLAN 配置信息：

```
[S3]disp vlan all
VLAN ID: 1
VLAN type: Static
Route interface: Not configured
Description: VLAN 0001
Name: VLAN 0001
Tagged ports:    None
Untagged ports:
   FortyGigE1/0/53            FortyGigE1/0/54
   GigabitEthernet1/0/21      GigabitEthernet1/0/23
   GigabitEthernet1/0/24      GigabitEthernet1/0/25
   GigabitEthernet1/0/26      GigabitEthernet1/0/27
   GigabitEthernet1/0/28      GigabitEthernet1/0/29
   GigabitEthernet1/0/30      GigabitEthernet1/0/31
   GigabitEthernet1/0/32      GigabitEthernet1/0/33
   GigabitEthernet1/0/34      GigabitEthernet1/0/35
   GigabitEthernet1/0/36      GigabitEthernet1/0/37
   GigabitEthernet1/0/38      GigabitEthernet1/0/39
   GigabitEthernet1/0/40      GigabitEthernet1/0/41
   GigabitEthernet1/0/42      GigabitEthernet1/0/43
   GigabitEthernet1/0/44      GigabitEthernet1/0/45
   GigabitEthernet1/0/46      GigabitEthernet1/0/47
   GigabitEthernet1/0/48
   Ten-GigabitEthernet1/0/49
```

```
    Ten-GigabitEthernet1/0/50
    Ten-GigabitEthernet1/0/51
    Ten-GigabitEthernet1/0/52

 VLAN ID: 10
 VLAN type: Static
 Route interface: Not configured
 Description: VLAN 0010
 Name: yewu1
 Tagged ports:
    GigabitEthernet1/0/21      GigabitEthernet1/0/23
    GigabitEthernet1/0/24               //允许通过的trunk接口
 Untagged ports:
    GigabitEthernet1/0/1       GigabitEthernet1/0/2
    GigabitEthernet1/0/3       GigabitEthernet1/0/4
    GigabitEthernet1/0/5

 VLAN ID: 20
 VLAN type: Static
 Route interface: Not configured
 Description: VLAN 0020
 Name: yewu2
 Tagged ports:
    GigabitEthernet1/0/21      GigabitEthernet1/0/23
    GigabitEthernet1/0/24               //允许通过的trunk接口
 Untagged ports:
    GigabitEthernet1/0/6       GigabitEthernet1/0/7
    GigabitEthernet1/0/8       GigabitEthernet1/0/9
    GigabitEthernet1/0/10

 VLAN ID: 30
 VLAN type: Static
 Route interface: Not configured
 Description: VLAN 0030
 Name: yewu3
 Tagged ports:
    GigabitEthernet1/0/21      GigabitEthernet1/0/23
    GigabitEthernet1/0/24               //允许通过的trunk接口
 Untagged ports:
    GigabitEthernet1/0/11      GigabitEthernet1/0/12
    GigabitEthernet1/0/13      GigabitEthernet1/0/14
    GigabitEthernet1/0/15

 VLAN ID: 40
 VLAN type: Static
```

```
Route interface: Not configured
Description: VLAN 0040
Name: yewu4
Tagged ports:
    GigabitEthernet1/0/21        GigabitEthernet1/0/23    //允许通过的trunk接口
Untagged ports:
    GigabitEthernet1/0/16        GigabitEthernet1/0/17
    GigabitEthernet1/0/18        GigabitEthernet1/0/19

VLAN ID: 50
VLAN type: Static
Route interface: Not configured
Description: VLAN 0050
Name: yewu5
Tagged ports:
    GigabitEthernet1/0/21        GigabitEthernet1/0/23    //允许通过的trunk接口
Untagged ports:
    GigabitEthernet1/0/20        GigabitEthernet1/0/22
```

⑤ 根据以上配置信息，整理出交换机 S2 和 S3 的 VLAN 配置信息和空闲接口信息，并将这些信息与项目规划进行对比。

2）根据相关知识点确定故障点的位置

分析交换机 S2 的 VLAN 配置信息，发现交换机 S2 的 G1/0/17 接口配置错误。在项目规划中，交换机 S2 的 G1/0/17 接口属于 VLAN 40，而从以上 VLAN 配置信息看，配置过程中把 S2 的 G1/0/17 接口划分到了 VLAN 50 中，这是一个故障点。同时，交换机 S2 上连接接口 trunk 接口中没有允许新增的 VLAN 通过。

由于交换机 S3 上的 PC 之间还存在故障点，因此继续分析交换机 S3 的 VLAN 配置信息。通过分析发现交换机 S3 上所有的 VLAN 配置信息和设计信息均相符，那故障点在什么地方呢？

通过进一步分析发现连接终端 PC_20 的接口为交换机 S3 的 G1/0/21 接口，而 S3 的 G1/0/21 接口的接口类型为 trunk 接口。结合 VLAN 原理来分析，只有将连接终端的接口类型设置为 access 接口才能互连终端设备，所以这是一个故障点。

通过以上分析，确定故障点主要有如下几个。

① 交换机 S2 的 G1/0/17 接口应属于 VLAN 40。在 G1/0/24 接口 trunk 干道上配置允许 VLAN 40 和 VLAN 50 通过。

② 交换机 S3 的 G1/0/21 接口的接口类型应为 access 接口。

③ 分别在交换机 S1 的 G1/0/23 和 G1/0/24 接口 trunk 干道配置允许 VLAN 40 和 VLAN 50 通过。

3）修改配置，保存配置信息并进行测试

① 修改配置。

进入交换机 S2：

```
[S2] vlan 40
[S2-vlan40]port GigabitEthernet 1/0/17
[S2-vlan40]exit
[S2]write changed                //保存配置信息
The current configuration will be written to the device. Are you sure? [Y/N]:y
Please input the file name(*.cfg)[flash:/startup.cfg]
(To leave the existing filename unchanged, press the enter key):
flash:/startup.cfg exists, overwrite? [Y/N]:y
Validating file. Please wait...
Saved the current configuration to mainboard device successfully.
[S2]interface ge 1/0/24
[S2-GigabitEthernet1/0/24]port trunk
[S2-GigabitEthernet1/0/24]port trunk permit vlan all
```

同理，对交换机 S1 的 G1/0/23 和 G1/0/24 接口做类似操作。

进入交换机 S3：

```
[S3]inter GigabitEthernet 1/0/21
[S3-GigabitEthernet1/0/21]no port link-type
[S3-GigabitEthernet1/0/21]exit
[S3]vlan 50
[S3-vlan50]port GigabitEthernet 1/0/21
[S3-vlan50]exit
[S3]write changed
The current configuration will be written to the device. Are you sure? [Y/N]:y
Please input the file name(*.cfg)[flash:/startup.cfg]
(To leave the existing filename unchanged, press the enter key):
flash:/startup.cfg exists, overwrite? [Y/N]:y
Validating file. Please wait...
Saved the current configuration to mainboard device successfully.
```

② 进行测试。新环境下的测试结果如表 3-5 所示。

表 3-5 新环境下的测试结果

测试序号	交换机 S2	交换机 S3	测试方法	预期测试结果	实际测试结果	是否发生故障
1	PC_13、PC_14		ping 命令	成功	成功	否
2	PC_15	PC_20	ping 命令	成功	成功	否
3	PC_16	PC_20	ping 命令	成功	成功	否

③ 整理新的配置文档。在故障排除后，保存所有交换机的配置信息，并更新书面的记录材料，确保书面文档和实际配置保持一致，以确保下次配置正常使用。

3.3 相关知识准备

为了能够深入地分析故障点，读者应了解 VLAN 的相关知识。

知识准备

1）什么是 VLAN

VLAN（Virtual LAN）翻译成中文是"虚拟局域网"。LAN 可以是由少数几台计算机构成的网络，也可以是由数以百计的计算机构成的企业网络。由于 LAN 内通过交换机连接的计算机过多，并且这些计算机又处于同一个广播域，因此既影响了网络性能，又带来了安全隐患。若采用不同交换机分离各部门的计算机，不仅投资较大，而且交换机得不到充分利用。行业内通常使用逻辑隔离的方式实现一个广播域的隔离，让同一台交换机上相连的计算机能够分别处于不同的逻辑 LAN，这就是 VLAN。

广播域是指广播帧（目标 MAC 地址全部为 1）所能传递的范围，也就是能够直接进行通信的范围。严格地说，除了广播帧，多播帧（Multicast Frame）和目标不明的单播帧（Unknown Unicast Frame）也能在同一个广播域中畅行无阻。

2）VLAN 传输过程

当交换式以太网出现后，同一个交换机不同的接口位于不同的冲突域中，从而大大提高了网络通信的效率。但是，在交换式以太网中，由于交换机所有的接口都位于同一个广播域中，导致局域网中的所有计算机都能够接收到一台计算机发出的广播帧，这会导致局域网中有限的网络资源被无用的广播信息占用。

在图 3-3 中，4 台主机发出的广播帧在整个局域网中广播。假如每台主机的广播帧流是 100Kbps，则 4 台主机的广播帧流可达到 400Kbps；如果链路带宽是 100Mbps，那么广播帧将占用 0.4%的带宽。但是，如果网络内主机达到 400 台，那么广播帧流将达到 40Mbps，广播帧将占用 40%的带宽，网络上到处充斥着广播帧流，这会严重浪费网络带宽资源。另外，过多的广播帧流会造成网络设备及主机的 CPU 负担过重，系统反应变慢甚至死机。

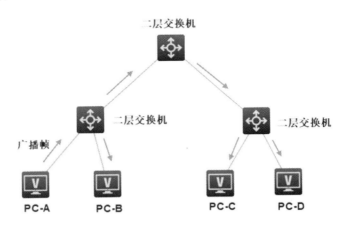

图 3-3 广播示意图

IEEE 专门制定了 802.1Q 协议标准来规定 VLAN 技术。利用 VLAN 技术可以实现在二层交换机上隔离广播域的目的，如图 3-4 所示。

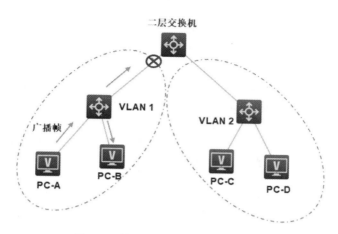

图 3-4 利用 VLAN 技术隔离广播域

以太网交换机根据 MAC 地址表来转发数据帧。MAC 地址表中包含了交换机接口和接口所连接终端主机的 MAC 地址映射关系。交换机从接口接收到以太网帧后,通过查找 MAC 地址表来决定从哪一个接口将以太网帧转发出去。若接口收到的是广播帧,则交换机会把广播帧从除原接口外的所有接口转发出去。

在 VLAN 技术中,通过给以太网帧添加一个标签(Tag)来标记这个以太网帧能够在哪个 VLAN 中传播。这样,交换机在转发数据帧时,不仅需要查找 MAC 地址表来决定将数据帧转发到哪个接口中,还需要检查接口上的 VLAN 标签是否匹配。

在图 3-5 中,交换机给主机 PCA 和 PCB 发送的以太网帧添加了 VLAN 10 的标签,给主机 PCC 和 PCD 发送的以太网帧添加了 VLAN 20 的标签,并在 MAC 地址表中增加关于 VLAN 标签的记录。这样,交换机在进行 MAC 地址表查找转发操作时会查看 VLAN 标签是否匹配,若不匹配,则交换机丢弃该数据帧。这样相当于用 VLAN 标签把 MAC 地址表中的表项区分开,只有相同 VLAN 标签的接口之间才能够互相转发数据帧。

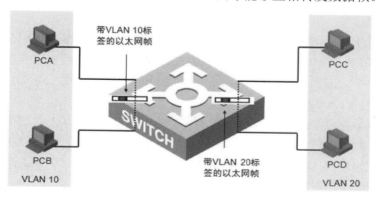

图 3-5 转发带 VLAN 标签的以太网帧

3)VLAN 帧格式

IEEE 802.1Q 协议标准规定了 VLAN 技术,其主要内容包括如下 3 部分。

① VLAN 的架构。
② VLAN 技术提供的服务。
③ VLAN 技术涉及的协议和算法。

为了保证不同厂家生产的设备能够顺利互通，IEEE 802.1Q 协议标准严格规定了统一的 VLAN 帧格式及其他重要参数。在此重点介绍标准的 VLAN 帧格式。

在传统的以太网帧中添加了 4 字节的 IEEE 802.1Q 标签的数据帧被称为带 VLAN 标签的帧（Tagged Frame），如图 3-6 所示；传统的不携带 IEEE 802.1Q 标签的数据帧被称为未打标签的帧（Untagged Frame）。

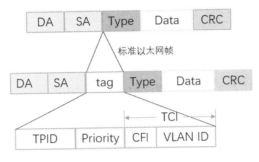

图 3-6 带 IEEE 802.1Q 标签的以太网帧

IEEE 802.1Q 标签头包含 2 字节的标签协议标识（Tag Protocol Identifier，TPID）和 2 字节的标签控制信息（Tag Control Information，TCI）。

TPID 是 IEEE 定义的新类型，表明这是一个封装了 IEEE 802.1Q 标签的帧。TPID 包含一个固定的值 0x8100。

TCI 包含的是帧的控制信息，它包含以下元素。
① Priority：占 3 位，指明数据帧的优先级。一共有 8 种优先级，用 0～7 表示。
② CFI（Canonical Format Indicator）：占 1 位，CFI 的值为 0 表示规范格式，值为 1 表示非规范格式。
③ VLAN ID（VLAN Identifier）：占 12 位，指明 VLAN 的编号。VLAN 编号有 4096 个，每个支持 IEEE 802.1Q 协议标准的交换机发送出来的数据帧都会包含 VLAN ID，以指明自己属于哪个 VLAN。

4）单交换机 VLAN 标签操作

交换机根据数据帧中的 VLAN 标签来判定数据帧属于哪个 VLAN，这个 VLAN 标签是由交换机接口在数据帧进入交换机时添加的。这样做的好处是 VLAN 对主机是透明的，主机不需要知道网络中 VLAN 是如何划分的，也不需要识别带 IEEE 802.1Q 标签的以太网帧，所有的相关事情由交换机负责。

如图 3-7 所示，当主机发出的以太网帧到达交换机接口时，交换机根据相关 VLAN 配置给进入接口的帧添加相应的 IEEE 802.1Q 标签。在默认情况下，所添加标签中的 VLAN ID 等于接口所属 VLAN 的 ID。接口所属的 VLAN 被称为接口默认 VLAN，又被称为 PVID（Port VLAN ID）。

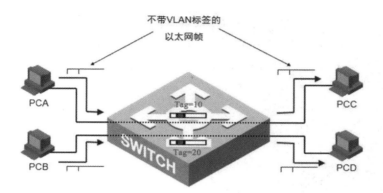

图 3-7 交换机 VLAN 标签操作

只允许默认 VLAN 的以太网帧通过的接口被称为 access 链路类型接口。access 链路类型接口在收到以太网帧后添加 VLAN 标签，在转发接口时剥离 VLAN 标签，所以对主机是透明的。

通常，在单交换机 VLAN 环境中，所有接口都是 access 链路类型接口。如图 3-8 所示，交换机连接 4 台终端主机，但这 4 台终端主机并不能识别带 VLAN 标签的以太网帧。通过在交换机上将与终端主机相连的接口设置为 access 链路类型接口，并指定该接口属于哪个 VLAN，使交换机能够根据接口进行 VLAN 划分，不同 VLAN 之间的接口属于不同广播域，从而隔离广播。

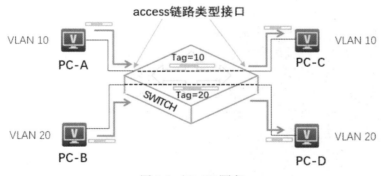

图 3-8 VLAN 隔离

5）VLAN 配置

完成一个 VLAN 配置主要包括以下几个步骤。

① 规划 VLAN 信息和主机 IP 地址。

② 设置 VLAN 信息，包括 VLAN ID 和 VLAN 命名。

③ 修改接口的模式信息。

④ 把接口添加到 VLAN 中。

6）VLAN 故障排除

依据结构化故障排除方法并结合 VLAN 的技术原理，故障排除可以分为以下几个任务。

① 查看原有交换机的配置，并按照要求填写表格。
② 查看新增 VLAN 后交换机的配置。
③ 根据相关知识点确定故障点的位置。
④ 修改错误配置，并保存配置信息。
⑤ 在新配置环境下进行测试。
⑥ 整理新的配置文档。

3.4 项目小结

本项目主要针对 VLAN 配置进行故障排除，介绍了当局域网出现故障时应该采用的故障排除方法和步骤，涉及的知识点主要包括 VLAN 的作用、原理、配置命令等。作为网络故障中最常见的故障，VLAN 故障的排除是网络管理员在学习网络故障排除的过程中要首先熟练掌握的内容。

素质拓展：测控保驾 神舟飞天

北京时间 2022 年 10 月 16 日 0 时 23 分，长征二号 F 运载火箭成功将神舟十三号载人飞船送入预定轨道，发射任务取得圆满成功。

神舟系列载人飞船是由多个系统组成的复杂系统。如何在保障飞行安全的同时，确保航天员的生命安全呢？这需要配套天地通信安控系统，包括多套地面安控系统、地面逃逸安控系统、车载逃逸安控系统和车载机动统一测控系统等关键安控、测控、逃逸任务。

统一测控系统的各测控站点需要遍布陆海空天、国内外，与多颗中继卫星共同组成立体通信测控网。测控系统精准助力，顺利完成交会"见面"，在有需要时对飞船进行测控，观测分析飞行器的位置、速度、飞行姿态，实时调节飞行姿态、轨迹、速度，实现精确定轨。整个测控通信网通过遥测、外测，确保精确控制航天器入轨。

天地通信系统：从神舟十二号发射任务开始，网络通信研究院陆续新建了多套固定站、车载站和便携站，并对卫星通信设备硬件及软件进行了升级改造。这保证了通信设备的高速稳定，实现了"超长待机"。

天地通信系统必须拥有大带宽通信能力，在天地之间打造高效、可靠的通信传输链路，可实现语音、视频、图像的双向传递，让航天员和地面人员进行实时交互。

苍穹浩瀚宇宙无垠，科学探索漫无止境，测控系统的建设者们肩负着为航天强国建设续写更大辉煌的使命，践行伟大航天精神，彰显大国重器担当，持续为中国航天"星辰大海"的梦想保驾护航。

增值服务

业务网络的日常维护需要工程技术人员具有全局意识和系统思维。在业务售后服务过程中，工程技术人员可以协助公司信息网络主管人员完善项目文档，厘清现有网络架构，注明设备在线运营清单，从而提升自身的日常运维能力及工作效率，为企业降低运维成本。

3.5 课后实训

项目内容：某企业现运行网络为星型拓扑结构，其核心交换机 S1 下连接了两台接入交换机 S2 和 S3，每台交换机上都存在 VLAN 10 和 VLAN 20 两个业务。最近由于业务扩展，扩容了 1 台接入交换机，因此网络管理员必须为新增的 1 台交换机创建新 VLAN 业务（VLAN 30），同时使用 192.168.10.0/24 网段分配局域网内的 PC 地址。考虑到业务需求，以后还可能会扩容接入交换机，因此采用 MVRP（Multiple VLAN Registration Protocol，多 VLAN 注册协议）。

在完成网络配置后进行测试时，网络出现了故障。请根据故障现象，完成以下任务。

① 根据要求检查故障现象。
② 根据故障现象收集故障信息。
③ 利用结构化故障排除方法完成故障定位。
④ 修改故障配置并说明故障原因。
⑤ 更新配置文档。

项目 4

局域网 STP 的故障排除

内容介绍

某厅局已完成局域网建设，局域网中主要包括以太网核心交换机（S1）和两台接入交换机（分别为 S2、S3），具体拓扑结构如图 4-1 所示。三台交换机之间存在网络冗余，采用 STP 确保网络的稳定性和可靠性。网络在投入使用一段时间后，为了监控各服务器的运行状态，网络管理员王工想要在局域网中添加一台网络管理机。

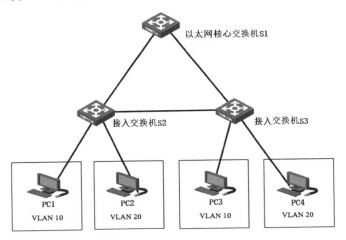

图 4-1　原有网络拓扑结构

任务安排

任务 1　针对新业务增加网络管理机

任务 2　进行网络更新过程中的故障分析与排除

项目 4　局域网 STP 的故障排除

学习目标

✧ 识别 STP 的常见故障
✧ 掌握故障排除的思路
✧ 学会结构化故障排除方法
✧ 学会 STP 相关故障排除及文档更新的方法

素质目标

乐于通过实践检验、判断各种技术问题，具有强烈的自我驱动意识，严谨细致地把工作做到最好。

4.1　STP 配置分析与实施

发现故障

某厅局已完成局域网建设，局域网中主要包括以太网核心交换机（S1）和两台接入交换机（分别为 S2、S3）。如图 4-1 所示，以太网核心交换机 S1 的 G0/1、G0/2 接口分别与接入交换机 S2 的 G0/1 接口、接入交换机 S3 的 G0/1 接口互连；PC1 和 PC2 分别接入 S2 的 G0/11 和 G0/12 接口，PC3 和 PC4 分别接入 S3 的 G0/11 和 G0/12 接口；PC1 和 PC3 属于 VLAN 10，PC2 和 PC4 属于 VLAN 20。从图 4-1 中可以看出，三台交换机之间存在网络冗余，形成了一条环路。为了避免环路造成网络故障，因此采用 STP 来确保网络的稳定性和可靠性。

网络在投入使用一段时间后，为了监控各服务器的运行状态，在局域网中添加了一台网络管理机（又被称为网管计算机）。网管计算机（PC5）通过接入交换机 S2 的 G0/3 接口接入局域网，并且通过以太网核心交换机 S1 的虚拟接口与其他设备进行互通。同时，网络管理员为网管计算机单独划分了一个 VLAN（VLAN 30）。新增网管计算机后的网络拓扑结构如图 4-2 所示。

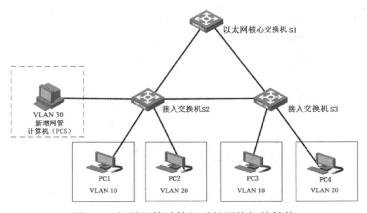

图 4-2　新增网管计算机后的网络拓扑结构

完成连接与配置后，对所有设备进行调试，发现新增网管计算机与其他设备存在互通故障。通过 ping 命令测试新增网管计算机与网络设备是否能够互通，结果如表 4-1 所示。

表 4-1 新增网管计算机与网络设备互通测试结果

测试序号	接入交换机 S2	接入交换机 S3	测试方法	预期测试结果	实际测试结果	是否发生故障
1	PC1、PC2、	PC3、PC4	ping 命令	成功	成功	否
2	新增网管计算机（PC5）	PC1、PC2	ping 命令	成功	失败	是
3	新增的网管计算机（PC5）	PC3、PC4	ping 命令	成功	失败	是

4.2 STP 配置故障分析与排除

1. 故障分析方法

根据结构化故障排除思路，可以将本次故障排除分解为以下几个任务。
① 查看各设备的地址配置是否存在问题。
② 查看各交换机的 VLAN 配置是否存在问题。
③ 根据相关知识点确定故障点的位置。
④ 修改错误配置，并保存配置信息。
⑤ 在新配置环境下进行测试。
⑥ 整理新的配置文档。

排除故障

2. 分析故障点

在发现故障现象后，网络管理员对新增网管计算机所涉及的设备和技术进行了详细分析，得出可能存在的故障点主要有以下几方面。

图 4-3 新增网管计算机的地址配置

① 从网络层分析，可能存在的故障点为地址配置错误。
② 从数据链路层分析，可能存在交换机的 VLAN 配置问题。
③ 从数据链路层分析，可能存在 VLAN 的数据流向与 STP 路径不一致的问题。
我们主要从上述分析进行故障排除。
1）查看各设备的地址配置
① 查看新增网管计算机的地址配置，如图 4-3 所示。
② 查看以太网核心交换机 S1 中的 VLAN 30 的 SVI 地址配置：

项目 4　局域网 STP 的故障排除

```
[H3C]disp ip int brief
*down: administratively down
(s): spoofing  (l): loopback
Interface              Physical        Protocol      IP Address       Description
MGE0/0/0               down            down          --               --
Vlan10                 up              up            192.168.10.254   --
Vlan20                 up              up            192.168.20.254   --
Vlan30                 up              up            192.168.30.254   --
```

③ 将所得结果填入表 4-2，并与规划 IP 地址进行比对。

表 4-2　现运行 IP 地址与规划 IP 地址比对

设备	现运行 IP 地址	规划 IP 地址	结果
S1-VLAN 30	192.168.30.254	192.168.30.254	正常
新增网管计算机（PC5）	192.168.30.5	192.168.30.5	正常

从表 4-2 可知，本次扩容不存在地址配置错误问题。接下来进行 VLAN 设置检查。

2）查看各交换机上的 VLAN 运行配置

① 查看以太网核心交换机 S1 的 VLAN 运行配置：

```
[S1]disp vlan all
VLAN ID: 1
VLAN type: Static
Route interface: Not configured
Description: VLAN 0001
Name: VLAN 0001
Tagged ports:   None
Untagged ports:
    FortyGigE1/0/53              FortyGigE1/0/54
    GigabitEthernet1/0/1         GigabitEthernet1/0/2
    GigabitEthernet1/0/3         GigabitEthernet1/0/4
    GigabitEthernet1/0/5         GigabitEthernet1/0/6
    GigabitEthernet1/0/7         GigabitEthernet1/0/8
    GigabitEthernet1/0/9         GigabitEthernet1/0/10
    GigabitEthernet1/0/11        GigabitEthernet1/0/12
    GigabitEthernet1/0/13        GigabitEthernet1/0/14
    GigabitEthernet1/0/15        GigabitEthernet1/0/16
    GigabitEthernet1/0/17        GigabitEthernet1/0/18
    GigabitEthernet1/0/19        GigabitEthernet1/0/20
    GigabitEthernet1/0/21        GigabitEthernet1/0/22
    GigabitEthernet1/0/23        GigabitEthernet1/0/24
    GigabitEthernet1/0/25        GigabitEthernet1/0/26
    GigabitEthernet1/0/27        GigabitEthernet1/0/28
    GigabitEthernet1/0/29        GigabitEthernet1/0/30
    GigabitEthernet1/0/31        GigabitEthernet1/0/32
    GigabitEthernet1/0/33        GigabitEthernet1/0/34
```

```
    GigabitEthernet1/0/35          GigabitEthernet1/0/36
    GigabitEthernet1/0/37          GigabitEthernet1/0/38
    GigabitEthernet1/0/39          GigabitEthernet1/0/40
    GigabitEthernet1/0/41          GigabitEthernet1/0/42
    GigabitEthernet1/0/43          GigabitEthernet1/0/44
    GigabitEthernet1/0/45          GigabitEthernet1/0/46
    GigabitEthernet1/0/47          GigabitEthernet1/0/48
    Ten-GigabitEthernet1/0/49
    Ten-GigabitEthernet1/0/50
    Ten-GigabitEthernet1/0/51
    Ten-GigabitEthernet1/0/52
VLAN ID: 10
VLAN type: Static
Route interface: Configured
IPv4 address: 192.168.10.254
IPv4 subnet mask: 255.255.255.0
Description: VLAN 0010
Name: VLAN 0010
Tagged ports:
    GigabitEthernet1/0/1           GigabitEthernet1/0/2
Untagged ports: None


VLAN ID: 20
VLAN type: Static
Route interface: Configured
IPv4 address: 192.168.20.254
IPv4 subnet mask: 255.255.255.0
Description: VLAN 0020
Name: VLAN 0020
Tagged ports:
    GigabitEthernet1/0/1           GigabitEthernet1/0/2
Untagged ports: None


VLAN ID: 30
VLAN type: Static
Route interface: Configured
IPv4 address: 192.168.30.254
IPv4 subnet mask: 255.255.255.0
Description: VLAN 0030
Name: VLAN 0030
Tagged ports:
    GigabitEthernet1/0/2                                   //允许通过的trunk接口
Untagged ports: None
```

② 查看接入交换机 S2 的 VLAN 运行配置：

```
[S2]disp vlan all
 VLAN ID: 1
 VLAN type: Static
 Route interface: Not configured
 Description: VLAN 0001
 Name: VLAN 0001
 Tagged ports:   None
 Untagged ports:
    FortyGigE1/0/53                FortyGigE1/0/54
    GigabitEthernet1/0/1           GigabitEthernet1/0/2
    GigabitEthernet1/0/4           GigabitEthernet1/0/5
    GigabitEthernet1/0/6           GigabitEthernet1/0/7
    GigabitEthernet1/0/8           GigabitEthernet1/0/9
    GigabitEthernet1/0/10          GigabitEthernet1/0/13
    GigabitEthernet1/0/14          GigabitEthernet1/0/15
    GigabitEthernet1/0/16          GigabitEthernet1/0/17
    GigabitEthernet1/0/18          GigabitEthernet1/0/19
    GigabitEthernet1/0/20          GigabitEthernet1/0/21
    GigabitEthernet1/0/22          GigabitEthernet1/0/23
    GigabitEthernet1/0/24          GigabitEthernet1/0/25
    GigabitEthernet1/0/26          GigabitEthernet1/0/27
    GigabitEthernet1/0/28          GigabitEthernet1/0/29
    GigabitEthernet1/0/30          GigabitEthernet1/0/31
    GigabitEthernet1/0/32          GigabitEthernet1/0/33
    GigabitEthernet1/0/34          GigabitEthernet1/0/35
    GigabitEthernet1/0/36          GigabitEthernet1/0/37
    GigabitEthernet1/0/38          GigabitEthernet1/0/39
    GigabitEthernet1/0/40          GigabitEthernet1/0/41
    GigabitEthernet1/0/42          GigabitEthernet1/0/43
    GigabitEthernet1/0/44          GigabitEthernet1/0/45
    GigabitEthernet1/0/46          GigabitEthernet1/0/47
    GigabitEthernet1/0/48
    Ten-GigabitEthernet1/0/49
    Ten-GigabitEthernet1/0/50
    Ten-GigabitEthernet1/0/51
    Ten-GigabitEthernet1/0/52

 VLAN ID: 10
 VLAN type: Static
 Route interface: Not configured
 Description: VLAN 0010
 Name: VLAN 0010
 Tagged ports:
    GigabitEthernet1/0/1           GigabitEthernet1/0/2
```

```
Untagged ports:
    GigabitEthernet1/0/11

VLAN ID: 20
VLAN type: Static
Route interface: Not configured
Description: VLAN 0020
Name: VLAN 0020
Tagged ports:
    GigabitEthernet1/0/1        GigabitEthernet1/0/2
Untagged ports:
    GigabitEthernet1/0/12

VLAN ID: 30
VLAN type: Static
Route interface: Not configured
Description: VLAN 0030
Name: VLAN 0030
Tagged ports:
    GigabitEthernet1/0/1        GigabitEthernet1/0/2
Untagged ports:
    GigabitEthernet1/0/3
```

③ 查看接入交换机 S3 的 VLAN 运行配置：

```
[S3] disp vlan all
VLAN ID: 1
VLAN type: Static
Route interface: Not configured
Description: VLAN 0001
Name: VLAN 0001
Tagged ports:   None
Untagged ports:
    FortyGigE1/0/53             FortyGigE1/0/54
    GigabitEthernet1/0/1        GigabitEthernet1/0/2
    GigabitEthernet1/0/3        GigabitEthernet1/0/4
    GigabitEthernet1/0/5        GigabitEthernet1/0/6
    GigabitEthernet1/0/7        GigabitEthernet1/0/8
    GigabitEthernet1/0/9        GigabitEthernet1/0/10
    GigabitEthernet1/0/13       GigabitEthernet1/0/14
    GigabitEthernet1/0/15       GigabitEthernet1/0/16
    GigabitEthernet1/0/17       GigabitEthernet1/0/18
    GigabitEthernet1/0/19       GigabitEthernet1/0/20
    GigabitEthernet1/0/21       GigabitEthernet1/0/22
    GigabitEthernet1/0/23       GigabitEthernet1/0/24
    GigabitEthernet1/0/25       GigabitEthernet1/0/26
```

```
    GigabitEthernet1/0/27       GigabitEthernet1/0/28
    GigabitEthernet1/0/29       GigabitEthernet1/0/30
    GigabitEthernet1/0/31       GigabitEthernet1/0/32
    GigabitEthernet1/0/33       GigabitEthernet1/0/34
    GigabitEthernet1/0/35       GigabitEthernet1/0/36
    GigabitEthernet1/0/37       GigabitEthernet1/0/38
    GigabitEthernet1/0/39       GigabitEthernet1/0/40
    GigabitEthernet1/0/41       GigabitEthernet1/0/42
    GigabitEthernet1/0/43       GigabitEthernet1/0/44
    GigabitEthernet1/0/45       GigabitEthernet1/0/46
    GigabitEthernet1/0/47       GigabitEthernet1/0/48
    Ten-GigabitEthernet1/0/49
    Ten-GigabitEthernet1/0/50
    Ten-GigabitEthernet1/0/51
    Ten-GigabitEthernet1/0/52

VLAN ID: 10
VLAN type: Static
Route interface: Not configured
Description: VLAN 0010
Name: VLAN 0010
Tagged ports:
    GigabitEthernet1/0/1        GigabitEthernet1/0/2
Untagged ports:
    GigabitEthernet1/0/11


VLAN ID: 20
VLAN type: Static
Route interface: Not configured
Description: VLAN 0020
Name: VLAN 0020
Tagged ports:
    GigabitEthernet1/0/1        GigabitEthernet1/0/2
Untagged ports:
    GigabitEthernet1/0/12


VLAN ID: 30
VLAN type: Static
Route interface: Not configured
Description: VLAN 0030
Name: VLAN 0030
Tagged ports:
    GigabitEthernet1/0/1        GigabitEthernet1/0/2        //允许通过的trunk接口
Untagged ports: None
```

通过查看各交换机的 VLAN 运行配置可以发现以太网核心交换机 S1 的 G1/0/1 接口为 trunk 接口，但没有设置允许 VLAN 30 通过。经分析可知，新增网管计算机的数据可以先通过接入交换机 S2 与接入交换机 S3 之间的 trunk 链路，再通过接入交换机 S3 与以太网核心交换机 S1 之间的 trunk 链路传到 S1 中。结合二层网络中的 STP，初步估计这里可能存在 VLAN 30 的数据与 STP 路径不一致的问题。接下来需要查看网络的 STP 运行状态。

3）查看网络的 STP 运行状态

① 查看以太网核心交换机 S1 的 STP 运行状态，注意各交换机 Tagged ports 的工作状态：

```
[S1]disp stp
-------[CIST Global Info][Mode MSTP]-------
 Bridge ID               : 0.96d2-3bcb-0100
 Bridge times            : Hello 2s MaxAge 20s FwdDelay 15s MaxHops 20
 Root ID/ERPC            : 0.96d2-3bcb-0100, 0
 RegRoot ID/IRPC         : 0.96d2-3bcb-0100, 0
 RootPort ID             : 0.0
 BPDU-Protection         : Disabled
 Bridge Config-
 Digest-Snooping         : Disabled
 TC or TCN received      : 6
 Time since last TC      : 0 days 4h:40m:28s

----[Port54(FortyGigE1/0/53)][DOWN]----
 Port protocol           : Enabled
 Port role               : Disabled Port
 Port ID                 : 128.54
 Port cost(Legacy)       : Config=auto, Active=200000
 Desg.bridge/port        : 0.96d2-3bcb-0100, 128.54
 Port edged              : Config=disabled, Active=disabled
 Point-to-Point          : Config=auto, Active=false
 Transmit limit          : 10 packets/hello-time
 TC-Restriction          : Disabled
 Role-Restriction        : Disabled
 Protection type         : Config=none, Active=none
 MST BPDU format         : Config=auto, Active=802.1s
 Port Config-
 Digest-Snooping         : Disabled
 Rapid transition        : False
 Num of VLANs mapped     : 1
 Port times              : Hello 2s MaxAge 20s FwdDelay 15s MsgAge 0s RemHops 20
 BPDU sent               : 0
       TCN: 0, Config: 0, RST: 0, MST: 0
 BPDU received           : 0
```

```
         TCN: 0, Config: 0, RST: 0, MST: 0

----[Port55(FortyGigE1/0/54)][DOWN]----
 Port protocol         : Enabled
 Port role             : Disabled Port
 Port ID               : 128.55
 Port cost(Legacy)     : Config=auto, Active=200000
 Desg.bridge/port      : 0.96d2-3bcb-0100, 128.55
 Port edged            : Config=disabled, Active=disabled
 Point-to-Point        : Config=auto, Active=false
 Transmit limit        : 10 packets/hello-time
 TC-Restriction        : Disabled
 Role-Restriction      : Disabled
 Protection type       : Config=none, Active=none
 MST BPDU format       : Config=auto, Active=802.1s
 Port Config-
 Digest-Snooping       : Disabled
 Rapid transition      : False
 Num of VLANs mapped : 1
 Port times            : Hello 2s MaxAge 20s FwdDelay 15s MsgAge 0s RemHops 20
 BPDU sent             : 0
         TCN: 0, Config: 0, RST: 0, MST: 0
 BPDU received         : 0
         TCN: 0, Config: 0, RST: 0, MST: 0

----[Port2(GigabitEthernet1/0/1)][FORWARDING]----              //转发状态
 Port protocol         : Enabled
 Port role             : Designated Port (Boundary)
 Port ID               : 128.2
 Port cost(Legacy)     : Config=auto, Active=20
 Desg.bridge/port      : 0.96d2-3bcb-0100, 128.2
 Port edged            : Config=disabled, Active=disabled
 Point-to-Point        : Config=auto, Active=true
 Transmit limit        : 10 packets/hello-time
 TC-Restriction        : Disabled
 Role-Restriction      : Disabled
 Protection type       : Config=none, Active=none
 MST BPDU format       : Config=auto, Active=802.1s
 Port Config-
 Digest-Snooping       : Disabled
 Rapid transition      : True
 Num of VLANs mapped : 3
 Port times            : Hello 2s MaxAge 20s FwdDelay 15s MsgAge 0s RemHops 20
 BPDU sent             : 13452
         TCN: 0, Config: 0, RST: 0, MST: 13452
```

```
 BPDU received         : 3
         TCN: 0, Config: 0, RST: 0, MST: 3

----[Port3(GigabitEthernet1/0/2)][FORWARDING]----          //转发状态
 Port protocol         : Enabled
 Port role             : Designated Port (Boundary)
 Port ID               : 128.3
 Port cost(Legacy)     : Config=auto, Active=20
 Desg.bridge/port      : 0.96d2-3bcb-0100, 128.3
 Port edged            : Config=disabled, Active=disabled
 Point-to-Point        : Config=auto, Active=true
 Transmit limit        : 10 packets/hello-time
 TC-Restriction        : Disabled
 Role-Restriction      : Disabled
 Protection type       : Config=none, Active=none
 MST BPDU format       : Config=auto, Active=802.1s
 Port Config-
 Digest-Snooping       : Disabled
 Rapid transition      : True
 Num of VLANs mapped   : 4
 Port times            : Hello 2s MaxAge 20s FwdDelay 15s MsgAge 0s RemHops 20
 BPDU sent             : 8415
         TCN: 0, Config: 0, RST: 0, MST: 8415
 BPDU received         : 1
         TCN: 0, Config: 0, RST: 0, MST: 1

----[Port4(GigabitEthernet1/0/3)][DOWN]----
 Port protocol         : Enabled
 Port role             : Disabled Port
 Port ID               : 128.4
 Port cost(Legacy)     : Config=auto, Active=200000
 Desg.bridge/port      : 0.96d2-3bcb-0100, 128.4
 Port edged            : Config=disabled, Active=disabled
 Point-to-Point        : Config=auto, Active=false
 Transmit limit        : 10 packets/hello-time
 TC-Restriction        : Disabled
 Role-Restriction      : Disabled
 Protection type       : Config=none, Active=none
 MST BPDU format       : Config=auto, Active=802.1s
 Port Config-
 Digest-Snooping       : Disabled
 Rapid transition      : False
 Num of VLANs mapped   : 1
 Port times            : Hello 2s MaxAge 20s FwdDelay 15s MsgAge 0s RemHops 20
 BPDU sent             : 0
```

```
         TCN: 0, Config: 0, RST: 0, MST: 0
 BPDU received        : 0
         TCN: 0, Config: 0, RST: 0, MST:

[S1]disp stp brief
 MST ID    Port                          Role   STP State    Protection
  0        GigabitEthernet1/0/1          DESI   FORWARDING   NONE
  0        GigabitEthernet1/0/2          DESI   FORWARDING   NONE
```

② 查看接入交换机 S2 的 STP 运行状态：

```
[S2]disp stp
-------[CIST Global Info][Mode MSTP]-------
 Bridge ID            : 4096.96d2-3fac-0200
 Bridge times         : Hello 2s MaxAge 20s FwdDelay 15s MaxHops 20
 Root ID/ERPC         : 0.96d2-3bcb-0100, 20
 RegRoot ID/IRPC      : 4096.96d2-3fac-0200, 0
 RootPort ID          : 128.2
 BPDU-Protection      : Disabled
 Bridge Config-
 Digest-Snooping      : Disabled
 TC or TCN received   : 8
 Time since last TC   : 0 days 4h:33m:11s

----[Port54(FortyGigE1/0/53)][DOWN]----
 Port protocol        : Enabled
 Port role            : Disabled Port
 Port ID              : 128.54
 Port cost(Legacy)    : Config=auto, Active=200000
 Desg.bridge/port     : 4096.96d2-3fac-0200, 128.54
 Port edged           : Config=disabled, Active=disabled
 Point-to-Point       : Config=auto, Active=false
 Transmit limit       : 10 packets/hello-time
 TC-Restriction       : Disabled
 Role-Restriction     : Disabled
 Protection type      : Config=none, Active=none
 MST BPDU format      : Config=auto, Active=802.1s
 Port Config-
 Digest-Snooping      : Disabled
 Rapid transition     : False
 Num of VLANs mapped  : 1
 Port times           : Hello 2s MaxAge 20s FwdDelay 15s MsgAge 0s RemHops 20
 BPDU sent            : 0
         TCN: 0, Config: 0, RST: 0, MST: 0
 BPDU received        : 0
         TCN: 0, Config: 0, RST: 0, MST: 0
```

```
----[Port55(FortyGigE1/0/54)][DOWN]----
 Port protocol         : Enabled
 Port role             : Disabled Port
 Port ID               : 128.55
 Port cost(Legacy)     : Config=auto, Active=200000
 Desg.bridge/port      : 4096.96d2-3fac-0200, 128.55
 Port edged            : Config=disabled, Active=disabled
 Point-to-Point        : Config=auto, Active=false
 Transmit limit        : 10 packets/hello-time
 TC-Restriction        : Disabled
 Role-Restriction      : Disabled
 Protection type       : Config=none, Active=none
 MST BPDU format       : Config=auto, Active=802.1s
 Port Config-
  Digest-Snooping      : Disabled
 Rapid transition      : False
 Num of VLANs mapped   : 1
 Port times            : Hello 2s MaxAge 20s FwdDelay 15s MsgAge 0s RemHops 20
 BPDU sent             : 0
        TCN: 0, Config: 0, RST: 0, MST: 0
 BPDU received         : 0
        TCN: 0, Config: 0, RST: 0, MST: 0

----[Port2(GigabitEthernet1/0/1)][FORWARDING]----           //转发状态
 Port protocol         : Enabled
 Port role             : Root Port (Boundary)
 Port ID               : 128.2
 Port cost(Legacy)     : Config=auto, Active=20
 Desg.bridge/port      : 0.96d2-3bcb-0100, 128.2
 Port edged            : Config=disabled, Active=disabled
 Point-to-Point        : Config=auto, Active=true
 Transmit limit        : 10 packets/hello-time
 TC-Restriction        : Disabled
 Role-Restriction      : Disabled
 Protection type       : Config=none, Active=none
 MST BPDU format       : Config=auto, Active=802.1s
 Port Config-
  Digest-Snooping      : Disabled
 Rapid transition      : True
 Num of VLANs mapped   : 4
 Port times            : Hello 2s MaxAge 20s FwdDelay 15s MsgAge 0s RemHops 20
 BPDU sent             : 3
        TCN: 0, Config: 0, RST: 0, MST: 3
 BPDU received         : 13631
        TCN: 0, Config: 0, RST: 0, MST: 13631
```

```
----[Port3(GigabitEthernet1/0/2)][DOWN]----    - //G1/0/2 接口为 DOWN
 Port protocol       : Enabled
 Port role           : Disabled Port
 Port ID             : 128.3
 Port cost(Legacy)   : Config=auto, Active=200000
 Desg.bridge/port    : 4096.96d2-3fac-0200, 128.3
 Port edged          : Config=disabled, Active=disabled
 Point-to-Point      : Config=auto, Active=false
 Transmit limit      : 10 packets/hello-time
 TC-Restriction      : Disabled
 Role-Restriction    : Disabled
 Protection type     : Config=none, Active=none
 MST BPDU format     : Config=auto, Active=802.1s
 Port Config-
 Digest-Snooping     : Disabled
 Rapid transition    : False
 Num of VLANs mapped : 4
 Port times          : Hello 2s MaxAge 20s FwdDelay 15s MsgAge 0s RemHops 20
 BPDU sent           : 0
        TCN: 0, Config: 0, RST: 0, MST: 0
 BPDU received       : 0
        TCN: 0, Config: 0, RST: 0, MST: 0

----[Port4(GigabitEthernet1/0/3)][FORWARDING]----
 Port protocol       : Enabled
 Port role           : Designated Port
 Port ID             : 128.4
 Port cost(Legacy)   : Config=auto, Active=20
 Desg.bridge/port    : 4096.96d2-3fac-0200, 128.4
 Port edged          : Config=enabled, Active=enabled
 Point-to-Point      : Config=auto, Active=true
 Transmit limit      : 10 packets/hello-time
 TC-Restriction      : Disabled
 Role-Restriction    : Disabled
 Protection type     : Config=none, Active=none
 MST BPDU format     : Config=auto, Active=802.1s
 Port Config-
 Digest-Snooping     : Disabled
 Rapid transition    : True
 Num of VLANs mapped : 1
 Port times          : Hello 2s MaxAge 20s FwdDelay 15s MsgAge 1s RemHops 20
 BPDU sent           : 14457
        TCN: 0, Config: 0, RST: 0, MST: 14457
 BPDU received       : 0
```

```
            TCN: 0, Config: 0, RST: 0, MST: 0
[S2]disp stp brief
 MST ID    Port                         Role  STP State   Protection
 0         GigabitEthernet1/0/1         ROOT  FORWARDING  NONE
 0         GigabitEthernet1/0/3         DESI  FORWARDING  NONE
 0         GigabitEthernet1/0/11        DESI  FORWARDING  NONE
 0         GigabitEthernet1/0/12        DESI  FORWARDING  NONE
```

③ 查看接入交换机 S3 的 STP 运行状态：

```
[S3]disp stp
-------[CIST Global Info][Mode MSTP]-------
Bridge ID              : 4096.96d2-41fb-0300
Bridge times           : Hello 2s MaxAge 20s FwdDelay 15s MaxHops 20
Root ID/ERPC           : 0.96d2-3bcb-0100, 20
RegRoot ID/IRPC        : 4096.96d2-41fb-0300, 0
RootPort ID            : 128.2
BPDU-Protection        : Disabled
Bridge Config-
Digest-Snooping        : Disabled
TC or TCN received     : 6
Time since last TC     : 0 days 4h:49m:44s

----[Port54(FortyGigE1/0/53)][DOWN]----
Port protocol          : Enabled
Port role              : Disabled Port
Port ID                : 128.54
Port cost(Legacy)      : Config=auto, Active=200000
Desg.bridge/port       : 4096.96d2-41fb-0300, 128.54
Port edged             : Config=disabled, Active=disabled
Point-to-Point         : Config=auto, Active=false
Transmit limit         : 10 packets/hello-time
TC-Restriction         : Disabled
Role-Restriction       : Disabled
Protection type        : Config=none, Active=none
MST BPDU format        : Config=auto, Active=802.1s
Port Config-
Digest-Snooping        : Disabled
Rapid transition       : False
Num of VLANs mapped    : 1
Port times             : Hello 2s MaxAge 20s FwdDelay 15s MsgAge 0s RemHops 20
BPDU sent              : 0
       TCN: 0, Config: 0, RST: 0, MST: 0
BPDU received          : 0
       TCN: 0, Config: 0, RST: 0, MST: 0
```

```
----[Port55(FortyGigE1/0/54)][DOWN]----
 Port protocol          : Enabled
 Port role              : Disabled Port
 Port ID                : 128.55
 Port cost(Legacy)      : Config=auto, Active=200000
 Desg.bridge/port       : 4096.96d2-41fb-0300, 128.55
 Port edged             : Config=disabled, Active=disabled
 Point-to-Point         : Config=auto, Active=false
 Transmit limit         : 10 packets/hello-time
 TC-Restriction         : Disabled
 Role-Restriction       : Disabled
 Protection type        : Config=none, Active=none
 MST BPDU format        : Config=auto, Active=802.1s
 Port Config-
 Digest-Snooping        : Disabled
 Rapid transition       : False
 Num of VLANs mapped    : 1
 Port times             : Hello 2s MaxAge 20s FwdDelay 15s MsgAge 0s RemHops 20
 BPDU sent              : 0
      TCN: 0, Config: 0, RST: 0, MST: 0
 BPDU received          : 0
      TCN: 0, Config: 0, RST: 0, MST: 0

----[Port2(GigabitEthernet1/0/1)][FORWARDING]----        //转发状态
 Port protocol          : Enabled
 Port role              : Root Port (Boundary)
 Port ID                : 128.2
 Port cost(Legacy)      : Config=auto, Active=20
 Desg.bridge/port       : 0.96d2-3bcb-0100, 128.3
 Port edged             : Config=disabled, Active=disabled
 Point-to-Point         : Config=auto, Active=true
 Transmit limit         : 10 packets/hello-time
 TC-Restriction         : Disabled
 Role-Restriction       : Disabled
 Protection type        : Config=none, Active=none
 MST BPDU format        : Config=auto, Active=802.1s
 Port Config-
 Digest-Snooping        : Disabled
 Rapid transition       : True
 Num of VLANs mapped    : 4
 Port times             : Hello 2s MaxAge 20s FwdDelay 15s MsgAge 0s RemHops 20
 BPDU sent              : 4
      TCN: 0, Config: 0, RST: 0, MST: 4
 BPDU received          : 15262
      TCN: 0, Config: 0, RST: 0, MST: 15262
```

```
----[Port3(GigabitEthernet1/0/2)][DOWN]----          //G1/0/2接口为DOWN
 Port protocol          : Enabled
 Port role              : Disabled Port
 Port ID                : 128.3
 Port cost(Legacy)      : Config=auto, Active=200000
 Desg.bridge/port       : 4096.96d2-41fb-0300, 128.3
 Port edged             : Config=disabled, Active=disabled
 Point-to-Point         : Config=auto, Active=false
 Transmit limit         : 10 packets/hello-time
 TC-Restriction         : Disabled
 Role-Restriction       : Disabled
 Protection type        : Config=none, Active=none
 MST BPDU format        : Config=auto, Active=802.1s
 Port Config-
 Digest-Snooping        : Disabled
 Rapid transition       : False
 Num of VLANs mapped    : 4
 Port times             : Hello 2s MaxAge 20s FwdDelay 15s MsgAge 0s RemHops 20
 BPDU sent              : 0
          TCN: 0, Config: 0, RST: 0, MST: 0
 BPDU received          : 0
          TCN: 0, Config: 0, RST: 0, MST: 0
----[Port12(GigabitEthernet1/0/11)][FORWARDING]----
 Port protocol          : Enabled
 Port role              : Designated Port
 Port ID                : 128.12
 Port cost(Legacy)      : Config=auto, Active=20
 Desg.bridge/port       : 4096.96d2-41fb-0300, 128.12
 Port edged             : Config=enabled, Active=enabled
 Point-to-Point         : Config=auto, Active=true
 Transmit limit         : 10 packets/hello-time
 TC-Restriction         : Disabled
 Role-Restriction       : Disabled
 Protection type        : Config=none, Active=none
 MST BPDU format        : Config=auto, Active=802.1s
 Port Config-
 Digest-Snooping        : Disabled
 Rapid transition       : True
 Num of VLANs mapped    : 1
 Port times             : Hello 2s MaxAge 20s FwdDelay 15s MsgAge 1s RemHops 20
 BPDU sent              : 15243
          TCN: 0, Config: 0, RST: 0, MST: 15243
 BPDU received          : 0
          TCN: 0, Config: 0, RST: 0, MST: 0
```

```
----[Port13(GigabitEthernet1/0/12)][FORWARDING]----
 Port protocol        : Enabled
 Port role            : Designated Port
 Port ID              : 128.13
 Port cost(Legacy)    : Config=auto, Active=20
 Desg.bridge/port     : 4096.96d2-41fb-0300, 128.13
 Port edged           : Config=enabled, Active=enabled
 Point-to-Point       : Config=auto, Active=true
 Transmit limit       : 10 packets/hello-time
 TC-Restriction       : Disabled
 Role-Restriction     : Disabled
 Protection type      : Config=none, Active=none
 MST BPDU format      : Config=auto, Active=802.1s
 Port Config-
 Digest-Snooping      : Disabled
 Rapid transition     : True
 Num of VLANs mapped  : 1
 Port times           : Hello 2s MaxAge 20s FwdDelay 15s MsgAge 1s RemHops 20
 BPDU sent            : 15256
         TCN: 0, Config: 0, RST: 0, MST: 15256
 BPDU received        : 0
         TCN: 0, Config: 0, RST: 0, MST: 0
[S3 ]disp stp brief
 MST ID   Port                          Role  STP State   Protection
 0        GigabitEthernet1/0/1          ROOT  FORWARDING  NONE
 0        GigabitEthernet1/0/11         DESI  FORWARDING  NONE
 0        GigabitEthernet1/0/12         DESI  FORWARDING  NONE
```

根据上面得出的各交换机的 STP 运行状态可知，经过 STP 计算，在 VLAN 30 中将接入交换机 S3 连接接入交换机 S2 的 G1/0/2 接口设置为阻塞接口。因此，VLAN 30（网管计算机）的数据流只能通过 S2 和以太网核心交换机 S1 之间的链路与其他设备进行互通。只需将 S1 的 G1/0/1 接口（trunk 接口）设置为允许 VLAN 30 通过，即可排除故障。

4）修改配置，保存配置信息并进行测试

① 修改配置，并保存配置信息。在以太网核心交换机 S1 和接入交换机 S3 上修改配置：

```
<S1>sys
System View: return to User View with Ctrl+Z.
[S1]int g1/0/1
[S1-GigabitEthernet1/0/1]undo port trunk permit vlan 10 20
[S1-GigabitEthernet1/0/1]port trunk permit vlan 10 20 30
[S1-GigabitEthernet1/0/1]exit
[S1]write

[S3]int g1/0/2
[S3-GigabitEthernet1/0/2]undo shutdown
[S3]write
```

② 在新配置环境下进行测试，结果如表 4-3 所示。

表 4-3 新配置环境下的测试结果

网管计算机	测试方法	服务器	测试结果
PC5	ping 命令	PC1	正常
PC5	ping 命令	PC2	正常
PC5	ping 命令	PC3	正常
PC5	ping 命令	PC4	正常

③ 整理新的配置文档。在故障排除后，保存所有交换机的配置信息，并更新书面的记录材料，确保书面文档与实际配置保持一致，以确保下次配置正常使用。

4.3 相关知识准备

知识准备

为了能够深入地分析故障点，读者应了解 STP 的相关知识。

在组建局域网的过程中，连通性是最基本的要求。一个局域网通常由多台交换机互联而成。在保证连通性的同时，要求网络具有高可靠性，就需要在网络中设置冗余链路。但是，这会引起广播风暴等一系列问题，所以需要保证网络中不存在路径环路。同时，为了最大限度地利用网络带宽，实现网络投资的最大化，需要充分利用网络中的冗余链路。高效地配置冗余链路可以提高网络的可靠性和带宽。

1）STP 的产生背景

① 广播风暴。在交换网络中，如果一台主机向网络发送一个广播包，那么交换机会向除发送广播包的接口外的所有接口发送这个广播包。当网络中存在环路时，交换机从一个接口向外发送的广播包会在另一个接口收到；由于交换机并不知道此广播包是自己发出的，因此它仍然会向外发送。这样网络中的广播包会越来越多，便形成了广播风暴。一方面，广播风暴会大量占用网络带宽；另一方面，主机需要对收到的广播包进行分析处理，这会占用大量的主机资源。

② 多帧复制。在存在冗余链路的交换网络中，当交换机刚刚启动且地址表中没有任何记录时，某主机会向同一网络的另一主机发送一个单播帧。这时，由于交换机还没有学习到任何地址，因此交换机 A 会将这个帧泛洪到所有接口。当冗余交换机 B 收到这个帧后，同样由于没有学习到任何地址，因此它也会将这个帧泛洪到所有接口。这样，目的主机会先后收到多个同样的帧，从而造成帧的重复接收。

③ MAC 地址表不稳定。交换机维护 MAC 地址表的原理是查看收到数据帧的源 MAC 地址。当交换网络中存在环路时，交换机可能在某一个接口收到来自主机的数据帧，于是 MAC 地址表中会添加一条该接口对应该主机的 MAC 地址的记录；由于存在冗余链路，在经过一段时间后，交换机另一个接口也会收到该主机的同一数据帧，于是 MAC 地址表中的记录被修改为另一个接口对应该主机的 MAC 地址。这种过程非常频繁，严重影响了 MAC 地址表的稳定性。

那么应该怎样解决以上问题呢？人们可能首先想到的是保证网络不存在物理上的环路。但是，当网络变得复杂时，要保证没有任何环路是很困难的。在许多对网络可靠性要求高的场景中，为了能够提供不间断的网络服务，采用物理环路的冗余备份是最常用的手段。所以，保证网络不存在环路是不现实的。

IEEE 提供了一个很好的解决办法，那就是 IEEE 802.1D 协议标准中规定的 STP（Spanning Tree Protocol，生成树协议）。该协议标准可以通过阻断网络中存在的冗余链路来消除网络中可能存在的路径环路，并且在当前活动（Active）路径发生故障时，可以通过激活被阻断的冗余备份链路来保证网络的连通性，保障业务的不间断服务。

STP 可用于在网络中建立树型拓扑，消除网络中的环路，并且可以通过一定的方法实现路径冗余。STP 适合于所有厂商的网络设备，不过这些网络设备在配置和功能强度上会有所差别，但是其原理和应用效果是一样的。

2）STP 的技术原理

STP 的技术原理是通过在交换机之间传递一种特殊的协议报文，即网桥协议数据单元（Bridge Protocol Data Unit，BPDU），以此来确定网络的拓扑结构。BPDU 有两种，分别是配置 BPDU（Configuration BPDU）和拓扑更改通知 BPDU（Topology Change Notification BPDU，TCNBPDU）。前者用于计算无环的生成树，后者用于在二层网络的拓扑结构发生变化时产生用来缩短 CAM（Content Addressable Memory）表项的刷新时间（由默认的 300s 缩短为 15s）。

STP 的基本思想是按照"树"的结构来构造网络拓扑，消除网络中的环路，避免因存在环路而造成广播风暴问题。"树"的根是一个被称为根桥的桥设备。根桥是由交换机或网桥的 BID（Bridge ID）确定的。BID 最小的设备被称为二层网络中的根桥。BID 是由网桥优先级和网桥 MAC 地址组成的，不同厂商设备的网桥优先级字节个数可能不同。"树"的根桥会定时发送配置 BPDU，而非根桥接收配置 BPDU，刷新最佳 BPDU 并转发。这里的最佳 BPDU 指的是当前根桥所发送的 BPDU。如果接收到了下级 BPDU（新接入的设备会发送 BPDU，但该设备的 BID 比当前根桥大），那么该设备将向新接入的设备发送自己存储的最佳 BPDU，以告知其当前网络中的根桥；如果接收到的 BPDU 更优，那么将重新计算生成树拓扑。新接入的设备会定时收到"树"中最佳 BPDU。当新接入的设备在距离上一次接收到最佳 BPDU 后经过最长寿命（Max Age，默认 20s）仍未接收到最佳 BPDU 时，接口将进入监听状态，该设备将产生 TCN BPDU，并将其从根端口转发出去；上级设备在接收到 TCN BPDU 后将先发送确认信息，再向它的上级设备发送 TCN BPDU。此过程持续到根桥为止。根桥在后续发送的配置 BPDU 中将携带标记，表明拓扑结构已发生变化。一旦网络中的所有设备都接收到这个标记，CAM 表项的刷新时间就会从 300s 缩短为 15s。整个收敛的时间为 50s 左右。

生成树算法：STP 运行时所使用的算法（STA）。生成树算法很复杂，但是其过程可以归纳为以下 3 个部分。

① 选择根网桥。
② 选择根端口。
③ 选择指定端口（又被称为转发端口）。

选择根网桥的依据是交换机的网桥优先级。网桥优先级是用来衡量网桥在生成树算法中优先级的十进制数，取值范围是 0~65 535，默认值是 32 768。BID 是由网桥优先级和网桥 MAC 地址组成的，共 8 字节。由于交换机的网桥优先级都是默认的，因此在根网桥的选择中比较的一般是网桥 MAC 地址的大小，选择 MAC 地址小的为根网桥。

3）STP 的功能介绍

STP 最主要的功能是避免局域网中的单点故障和网络环路，解决环路连接带来的广播风暴问题。在某种意义上，STP 是一种网络保护技术，可以消除由于失误或意外带来的环路连接。STP 提供了为网络备份连接的可能性，可与 SDH 保护结合使用，构成以太环网的双重保护。新型以太单板支持符合 IEEE 802.1d 协议标准的 STP 及 IEEE 802.1w 协议标准的 RSTP（快速生成树协议），收敛速度可达到 1s。

4）STP 的不足

① 拓扑收敛慢。当网络拓扑结构发生改变时，STP 需要 50s 才能完成拓扑收敛。
② 不能提供负载均衡的功能。当网络中出现环路时，STP 只是简单地将环路进行阻塞（Block），这样会使该链路不能转发数据包，从而浪费网络资源。

4.4 项目小结

本项目主要针对数据链路层进行故障排除。虽然 STP 在数据链路层起到了关键的二层冗余作用，但是我们必须注意 STP 生成树与实际流量是否吻合。

素质拓展：安全之道 始于筑基

詹天佑曾说：学术虽精，道德不足，犹诸筑高屋于流沙之上，稍有震摇，无不倾倒。

随着网络应用与业务的深入融合，网络安全日益重要。2014 年 2 月 27 日，中央网络安全和信息化领导小组成立，中共中央总书记、国家主席、中央军委主席习近平担任组长，提出了"没有网络安全，就没有国家安全；没有信息化，就没有现代化。"自 2016 年以来，我国相继发布了《互联网域名管理办法》《中华人民共和国网络安全法》《互联网信息服务管理办法》《中华人民共和国数据安全法》《中华人民共和国个人信息保护法》《数据出境安全评估办法》等。2022 年，工业和信息化部编制了《工业和信息化部行政执法事项清单（2022 年版）》，新增了 15 条（第 247~261 条）涉及数据安全的行政执法事项，对违反数据安全相关法规的行为执行行政处罚。

增值服务

我们只有确认故障、分析故障从而思考增值服务，才能在业务上精益求精。例如，对于接入 VLAN 数量较多的应用场景，考虑使用 Private VLAN 来优化网络，从而减少汇聚交换机的 IP 地址量，解决通信安全问题和避免浪费 IP 地址。针对 STP 的原理，考虑多 VLAN 下的网络负载均衡，为用户提供 MSTP（Multiple Spanning Tree Protocol，多生成树协议）功能建议，培养质量意识和保持高效可靠的团队化合作精神。

4.5 课后实训

项目内容：某公司搬迁后需要组建新的局域网，新局域网中的网络设备主要由两台核心交换机（S1 与 S2）及两台接入交换机（S3 与 S4）构成。为确保冗余，两台接入交换机分别连接至两台核心交换机，而两台核心交换机之间也进行了互联。每台接入交换机上规划了 4 个 VLAN（VLAN 10、VLAN 20、VLAN 30 和 VLAN 40），并且分别使用网段 192.168.10.0/24、192.168.20.0/24、192.168.30.0/24 和 192.168.40.0/24 为 4 个 VLAN 划分地址（网关地址分别为 192.168.10.254、192.168.20.254、192.168.30.254 和 192.168.40.254）。VLAN 10 和 VLAN 20 的网关部署在核心交换机 S1 上，VLAN 30 和 VLAN 40 的网关部署在核心交换机 S2 上。要求各 VLAN 内部的终端可以互联互通，VLAN 10 与 VLAN 20 之间的终端可以互联互通，VLAN 30 和 VLAN 40 之间的终端可以互联互通。

在完成配置后，网络出现了故障现象。请回顾所学知识点完成下面各项任务。

① 根据要求检查故障现象。
② 根据故障现象收集故障信息。
③ 利用结构化故障排除方法完成故障定位。
④ 修改故障配置并说明故障原因。
⑤ 更新配置文档。

项目 5

交换机 VLAN 间路由故障排除

内容介绍

某市财政局的局域网系统最近有新的业务需要接入,所以要在原有交换机上完成新增 VLAN 配置,并确保新增业务与原有业务之间能够互联互通。为了实现上述目标,需要网络管理员完成配置更新操作。

任务安排

任务 1　针对新业务增加 VLAN
任务 2　进行新增业务与原有业务互联互通的故障分析与排除

学习目标

- ◇ 识别 SVI 的功能
- ◇ 掌握 SVI 数据转发的故障排除思路
- ◇ 学会结构化故障排除方法
- ◇ 学会 VLAN 间数据交换相关故障排除及文档更新的方法

素质目标

工作主动,耐心细致,有责任心,具备良好的沟通技巧和团队合作精神。

5.1 VLAN 间路由配置分析与实施

发现故障

某市财政局的网络管理员李工接到任务，要在局域网原有交换机上新增 VLAN 配置，因此需要完成如下配置更新操作。

① 查看原有交换机的配置，落实已有的 VLAN 和 IP 地址信息。
② 更新原有交换机的配置，包括终端互连交换机和通信中 VLAN 转发的交换机。
③ 增加新业务 SVI 地址，进行新增终端的互联互通测试，并分析测试过程中数据包的转发过程。

（1）翻阅网络工程竣工配置文档，查看原有交换机的配置，确认已有的 VLAN 和 IP 地址信息。

（2）按新增业务需求更新网络交换机的配置。

查阅网络工程竣工配置文档，原有网络拓扑结构如图 5-1 所示。该网络由 3 台交换机（S1、S2、S3）组成，在 S2、S3 上分别有两个 VLAN（VLAN 10、VLAN 20），VLAN 10 和 VLAN 20 通过 S1 进行路由的互联互通。

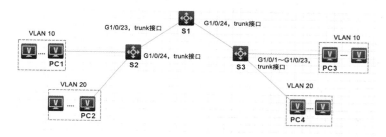

图 5-1 原有网络拓扑结构

在更新设备配置后，形成的新网络拓扑结构如图 5-2 所示。相比原来，S2 和 S3 上均增加了 VLAN 30，VLAN 30 同样利用 S1 进行路由的互联互通。在新的网络拓扑结构中，核心交换机 S1（三层交换机）的 G1/0/23 接口与 S2（二层交换机）的 G1/0/24 接口相连，G1/0/24 接口与 S3（二层交换机）的 G1/0/24 接口相连；终端 PC1、PC2、PC5 分别与 S2 的 G1/0/1 接口、G1/0/9 接口、G1/0/17 接口相连，终端 PC3、PC4、PC6 分别与 S3 的 G1/0/1 接口、G1/0/9 接口、G1/0/17 接口相连。

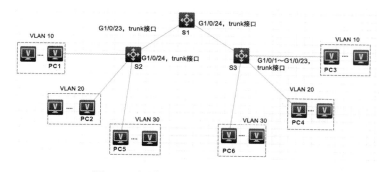

图 5-2 新增 VLAN 后的网络拓扑结构

（3）配置新业务 SVI 和终端 IP 地址并进行测试。

在完成网络配置后，对新增终端进行测试，结果如表 5-1 所示。

表 5-1 新增终端后的测试结果

测试序号	交换机 S2	交换机 S3	测试方法	预期测试结果	实际测试结果	是否发生故障
1	PC1、PC5		ping 命令	成功	失败	是
2	PC2、PC5		ping 命令	成功	失败	是
3		PC3、PC6	ping 命令	成功	失败	是
4		PC4、PC6	ping 命令	成功	失败	是
5	PC5	PC6	ping 命令	成功	失败	是

由表 5-1 可知，5 项测试均失败，因此本次操作没有成功完成项目目标。是什么原因造成的故障现象呢？是规划设计的问题、操作的问题，还是 VLAN 的 SVI 没有理解清楚的问题？据此，要深入进行故障分析，以确定问题的根源。

5.2 VLAN 间路由配置故障分析与排除

排除故障

1. 故障分析方法

依据结构化故障排除思路，结合 VLAN 间路由的技术原理，可以将本次故障排除分解为以下几个任务。

① 查看新增配置后各交换机的配置。

② 根据相关知识点确定同一 VLAN 内通信故障点的位置。

③ 根据相关知识点确定不同 VLAN 间通信故障点的位置。

④ 修改错误配置，并保存配置信息。

⑤ 在新配置环境下进行测试。

⑥ 整理新的配置文档。

2. 分析故障点

在发现故障现象后，网络管理员对本次任务所涉及的设备和技术进行了详细分析，得出可能存在的故障点主要有以下两个方面。

① 从网络层分析，可能存在的故障点为地址配置错误。

② 从数据链路层分析，可能存在交换机的 VLAN 配置问题。

我们主要从上述分析进行故障排除。

完成这次故障排除任务所需的设备如下。

① 一台装有超级终端软件或 Telnet 软件的计算机，同时确定访问所需的用户名和口令。

② 配置线缆。

③ 笔和纸，用于记录相关信息。

1）查看新增配置后各交换机的 VLAN 配置

① 使用超级终端软件或 Telnet 软件连接交换机，并使用 disp cu 命令查看交换机的 VLAN 配置。

查看 S1 的 VLAN 配置：

```
<S1>disp cu
#
 version 7.1.075, Alpha 7571
#
 sysname S1
#
 irf mac-address persistent timer
 irf auto-update enable
 undo irf link-delay
 irf member 1 priority 1
#
 lldp global enable
#
 system-working-mode standard
 xbar load-single
 password-recovery enable
 lpu-type f-series
#
vlan 1
#
vlan 10
 name yewu 10
#
vlan 20
 name yewu 20
#
vlan 30
 name yewu 30
#
 stp global enable
#
interface NULL0
#
interface Vlan-interface10
 ip address 192.168.10.254 255.255.255.0
#
interface Vlan-interface20
 ip address 192.168.20.254 255.255.255.0
#
interface Vlan-interface30
```

```
 ip address 192.168.30.254 255.255.255.0
#
interface FortyGigE1/0/53
 port link-mode bridge
#
interface FortyGigE1/0/54
 port link-mode bridge
#
interface GigabitEthernet1/0/1
 port link-mode bridge
 combo enable fiber
#
interface GigabitEthernet1/0/2
 port link-mode bridge
 combo enable fiber
#
interface GigabitEthernet1/0/3
 port link-mode bridge
 combo enable fiber
#
interface GigabitEthernet1/0/4
 port link-mode bridge
 combo enable fiber
#
interface GigabitEthernet1/0/5
 port link-mode bridge
 combo enable fiber
#
interface GigabitEthernet1/0/6
 port link-mode bridge
 combo enable fiber
#
interface GigabitEthernet1/0/7
 port link-mode bridge
 combo enable fiber
#
interface GigabitEthernet1/0/8
 port link-mode bridge
 combo enable fiber
#
interface GigabitEthernet1/0/9
 port link-mode bridge
 combo enable fiber
#
interface GigabitEthernet1/0/10
```

```
 port link-mode bridge
 combo enable fiber
#
interface GigabitEthernet1/0/11
 port link-mode bridge
 combo enable fiber
#
interface GigabitEthernet1/0/12
 port link-mode bridge
 combo enable fiber
#
interface GigabitEthernet1/0/13
 port link-mode bridge
 combo enable fiber
#
interface GigabitEthernet1/0/14
 port link-mode bridge
 combo enable fiber
#
interface GigabitEthernet1/0/15
 port link-mode bridge
 combo enable fiber
#
interface GigabitEthernet1/0/16
 port link-mode bridge
 combo enable fiber
#
interface GigabitEthernet1/0/17
 port link-mode bridge
 combo enable fiber
#
interface GigabitEthernet1/0/18
 port link-mode bridge
 combo enable fiber
#
interface GigabitEthernet1/0/19
 port link-mode bridge
 combo enable fiber
#
interface GigabitEthernet1/0/20
 port link-mode bridge
 combo enable fiber
#
interface GigabitEthernet1/0/21
 port link-mode bridge
```

```
 combo enable fiber
#
interface GigabitEthernet1/0/22
 port link-mode bridge
 combo enable fiber
#
interface GigabitEthernet1/0/23
 port link-mode bridge
 port link-type trunk
 port trunk permit vlan 1 10 20
 combo enable fiber
#
interface GigabitEthernet1/0/24
 port link-mode bridge
 port link-type trunk
 port trunk permit vlan 1 10 20
 combo enable fiber
#
interface GigabitEthernet1/0/25
 port link-mode bridge
 combo enable fiber
#
interface GigabitEthernet1/0/26
 port link-mode bridge
 combo enable fiber
#
interface GigabitEthernet1/0/27
 port link-mode bridge
 combo enable fiber
#
interface GigabitEthernet1/0/28
 port link-mode bridge
 combo enable fiber
#
interface GigabitEthernet1/0/29
 port link-mode bridge
 combo enable fiber
#
interface GigabitEthernet1/0/30
 port link-mode bridge
 combo enable fiber
#
interface GigabitEthernet1/0/31
 port link-mode bridge
 combo enable fiber
```

```
#
interface GigabitEthernet1/0/32
 port link-mode bridge
 combo enable fiber
#
interface GigabitEthernet1/0/33
 port link-mode bridge
 combo enable fiber
#
interface GigabitEthernet1/0/34
 port link-mode bridge
 combo enable fiber
#
interface GigabitEthernet1/0/35
 port link-mode bridge
 combo enable fiber
#
interface GigabitEthernet1/0/36
 port link-mode bridge
 combo enable fiber
#
interface GigabitEthernet1/0/37
 port link-mode bridge
 combo enable fiber
#
interface GigabitEthernet1/0/38
 port link-mode bridge
 combo enable fiber
#
interface GigabitEthernet1/0/39
 port link-mode bridge
 combo enable fiber
#
interface GigabitEthernet1/0/40
 port link-mode bridge
 combo enable fiber
#
interface GigabitEthernet1/0/41
 port link-mode bridge
 combo enable fiber
#
interface GigabitEthernet1/0/42
 port link-mode bridge
 combo enable fiber
#
```

```
interface GigabitEthernet1/0/43
 port link-mode bridge
 combo enable fiber
#
interface GigabitEthernet1/0/44
 port link-mode bridge
 combo enable fiber
#
interface GigabitEthernet1/0/45
 port link-mode bridge
 combo enable fiber
#
interface GigabitEthernet1/0/46
 port link-mode bridge
 combo enable fiber
#
interface GigabitEthernet1/0/47
 port link-mode bridge
 combo enable fiber
#
interface GigabitEthernet1/0/48
 port link-mode bridge
 combo enable fiber
#
interface M-GigabitEthernet0/0/0
#
interface Ten-GigabitEthernet1/0/49
 port link-mode bridge
 combo enable fiber
#
interface Ten-GigabitEthernet1/0/50
 port link-mode bridge
 combo enable fiber
#
interface Ten-GigabitEthernet1/0/51
 port link-mode bridge
 combo enable fiber
#
interface Ten-GigabitEthernet1/0/52
 port link-mode bridge
 combo enable fiber
#
 scheduler logfile size 16
#
line class aux
```

```
 user-role network-operator
#
line class console
 user-role network-admin
#
line class tty
 user-role network-operator
#
line class vty
 user-role network-operator
#
line aux 0
 user-role network-operator
#
line con 0
 user-role network-admin
#
line vty 0 63
 user-role network-operator
#
radius scheme system
 user-name-format without-domain
#
domain name system
#
 domain default enable system
#
role name level-0
 description Predefined level-0 role
#
role name level-1
 description Predefined level-1 role
#
role name level-2
 description Predefined level-2 role
#
role name level-3
 description Predefined level-3 role
#
role name level-4
 description Predefined level-4 role
#
role name level-5
 description Predefined level-5 role
#
```

```
role name level-6
 description Predefined level-6 role
#
role name level-7
 description Predefined level-7 role
#
role name level-8
 description Predefined level-8 role
#
role name level-9
 description Predefined level-9 role
#
role name level-10
 description Predefined level-10 role
#
role name level-11
 description Predefined level-11 role
#
role name level-12
 description Predefined level-12 role
#
role name level-13
 description Predefined level-13 role
#
role name level-14
 description Predefined level-14 role
#
user-group system
#
return
```

查看 S2 的 VLAN 配置：

```
<S2>disp cu
#
 version 7.1.075, Alpha 7571
#
 sysname S2
#
 irf mac-address persistent timer
 irf auto-update enable
 undo irf link-delay
 irf member 1 priority 1
#
 lldp global enable
#
 system-working-mode standard
```

```
 xbar load-single
 password-recovery enable
 lpu-type f-series
#
vlan 1
#
vlan 10
 name yewu10
#
vlan 20
 name yewu20
#
vlan 30
 name yewu30
#
 stp global enable
#
interface NULL0
#
interface FortyGigE1/0/53
 port link-mode bridge
#
interface FortyGigE1/0/54
 port link-mode bridge
#
interface GigabitEthernet1/0/1
 port link-mode bridge
 port access vlan 10
 combo enable fiber
#
interface GigabitEthernet1/0/2
 port link-mode bridge
 port access vlan 10
 combo enable fiber
#
interface GigabitEthernet1/0/3
 port link-mode bridge
 port access vlan 10
 combo enable fiber
#
interface GigabitEthernet1/0/4
 port link-mode bridge
 port access vlan 10
 combo enable fiber
#
```

```
interface GigabitEthernet1/0/5
 port link-mode bridge
 port access vlan 10
 combo enable fiber
#
interface GigabitEthernet1/0/6
 port link-mode bridge
 port access vlan 10
 combo enable fiber
#
interface GigabitEthernet1/0/7
 port link-mode bridge
 port access vlan 10
 combo enable fiber
#
interface GigabitEthernet1/0/8
 port link-mode bridge
 port access vlan 10
 combo enable fiber
#
interface GigabitEthernet1/0/9
 port link-mode bridge
 port access vlan 20
 combo enable fiber
#
interface GigabitEthernet1/0/10
 port link-mode bridge
 port access vlan 20
 combo enable fiber
#
interface GigabitEthernet1/0/11
 port link-mode bridge
 port access vlan 20
 combo enable fiber
#
interface GigabitEthernet1/0/12
 port link-mode bridge
 port access vlan 20
 combo enable fiber
#
interface GigabitEthernet1/0/13
 port link-mode bridge
 port access vlan 20
 combo enable fiber
#
```

```
interface GigabitEthernet1/0/14
 port link-mode bridge
 port access vlan 20
 combo enable fiber
#
interface GigabitEthernet1/0/15
 port link-mode bridge
 port access vlan 20
 combo enable fiber
#
interface GigabitEthernet1/0/16
 port link-mode bridge
 port access vlan 20
 combo enable fiber
#
interface GigabitEthernet1/0/17
 port link-mode bridge
 port access vlan 20
 combo enable fiber
#
interface GigabitEthernet1/0/18
 port link-mode bridge
 port access vlan 30
 combo enable fiber
#
interface GigabitEthernet1/0/19
 port link-mode bridge
 port access vlan 30
 combo enable fiber
#
interface GigabitEthernet1/0/20
 port link-mode bridge
 port access vlan 30
 combo enable fiber
#
interface GigabitEthernet1/0/21
 port link-mode bridge
 port access vlan 30
 combo enable fiber
#
interface GigabitEthernet1/0/22
 port link-mode bridge
 port access vlan 30
 combo enable fiber
#
```

```
interface GigabitEthernet1/0/23
 port link-mode bridge
 port access vlan 30
 combo enable fiber
#
interface GigabitEthernet1/0/24
 port link-mode bridge
 port link-type trunk
 port trunk permit vlan 1 10 20
 combo enable fiber
#
interface GigabitEthernet1/0/25
 port link-mode bridge
 combo enable fiber
#
interface GigabitEthernet1/0/26
 port link-mode bridge
 combo enable fiber
#
interface GigabitEthernet1/0/27
 port link-mode bridge
 combo enable fiber
#
interface GigabitEthernet1/0/28
 port link-mode bridge
 combo enable fiber
#
interface GigabitEthernet1/0/29
 port link-mode bridge
 combo enable fiber
#
interface GigabitEthernet1/0/30
 port link-mode bridge
 combo enable fiber
#
interface GigabitEthernet1/0/31
 port link-mode bridge
 combo enable fiber
#
interface GigabitEthernet1/0/32
 port link-mode bridge
 combo enable fiber
#
interface GigabitEthernet1/0/33
 port link-mode bridge
```

```
 combo enable fiber
#
interface GigabitEthernet1/0/34
 port link-mode bridge
 combo enable fiber
#
interface GigabitEthernet1/0/35
 port link-mode bridge
 combo enable fiber
#
interface GigabitEthernet1/0/36
 port link-mode bridge
 combo enable fiber
#
interface GigabitEthernet1/0/37
 port link-mode bridge
 combo enable fiber
#
interface GigabitEthernet1/0/38
 port link-mode bridge
 combo enable fiber
#
interface GigabitEthernet1/0/39
 port link-mode bridge
 combo enable fiber
#
interface GigabitEthernet1/0/40
 port link-mode bridge
 combo enable fiber
#
interface GigabitEthernet1/0/41
 port link-mode bridge
 combo enable fiber
#
interface GigabitEthernet1/0/42
 port link-mode bridge
 combo enable fiber
#
interface GigabitEthernet1/0/43
 port link-mode bridge
 combo enable fiber
#
interface GigabitEthernet1/0/44
 port link-mode bridge
 combo enable fiber
```

```
#
interface GigabitEthernet1/0/45
 port link-mode bridge
 combo enable fiber
#
interface GigabitEthernet1/0/46
 port link-mode bridge
 combo enable fiber
#
interface GigabitEthernet1/0/47
 port link-mode bridge
 combo enable fiber
#
interface GigabitEthernet1/0/48
 port link-mode bridge
 combo enable fiber
#
interface M-GigabitEthernet0/0/0
#
interface Ten-GigabitEthernet1/0/49
 port link-mode bridge
 combo enable fiber
#
interface Ten-GigabitEthernet1/0/50
 port link-mode bridge
 combo enable fiber
#
interface Ten-GigabitEthernet1/0/51
 port link-mode bridge
 combo enable fiber
#
interface Ten-GigabitEthernet1/0/52
 port link-mode bridge
 combo enable fiber
#
 scheduler logfile size 16
#
line class aux
 user-role network-operator
#
line class console
 user-role network-admin
#
line class tty
 user-role network-operator
```

```
#
line class vty
 user-role network-operator
#
line aux 0
 user-role network-operator
#
line con 0
 user-role network-admin
#
line vty 0 63
 user-role network-operator
#
radius scheme system
 user-name-format without-domain
#
domain name system
#
 domain default enable system
#
role name level-0
 description Predefined level-0 role
#
role name level-1
 description Predefined level-1 role
#
role name level-2
 description Predefined level-2 role
#
role name level-3
 description Predefined level-3 role
#
role name level-4
 description Predefined level-4 role
#
role name level-5
 description Predefined level-5 role
#
role name level-6
 description Predefined level-6 role
#
role name level-7
 description Predefined level-7 role
#
role name level-8
```

```
 description Predefined level-8 role
#
role name level-9
 description Predefined level-9 role
#
role name level-10
 description Predefined level-10 role
#
role name level-11
 description Predefined level-11 role
#
role name level-12
 description Predefined level-12 role
#
role name level-13
 description Predefined level-13 role
#
role name level-14
 description Predefined level-14 role
#
user-group system
#
return
```

查看 S3 的 VLAN 配置：

```
<S3>disp cu
#
 version 7.1.075, Alpha 7571
#
 sysname S3
#
 irf mac-address persistent timer
 irf auto-update enable
 undo irf link-delay
 irf member 1 priority 1
#
 lldp global enable
#
 system-working-mode standard
 xbar load-single
 password-recovery enable
 lpu-type f-series
#
vlan 1
#
vlan 10
```

```
 name yewu10
#
vlan 20
 name yewu20
#
vlan 30
 name yewu30
#
 stp global enable
#
interface NULL0
#
interface FortyGigE1/0/53
 port link-mode bridge
#
interface FortyGigE1/0/54
 port link-mode bridge
#
interface GigabitEthernet1/0/1
 port link-mode bridge
 port access vlan 10
 combo enable fiber
#
interface GigabitEthernet1/0/2
 port link-mode bridge
 port access vlan 10
 combo enable fiber
#
interface GigabitEthernet1/0/3
 port link-mode bridge
 port access vlan 10
 combo enable fiber
#
interface GigabitEthernet1/0/4
 port link-mode bridge
 port access vlan 10
 combo enable fiber
#
interface GigabitEthernet1/0/5
 port link-mode bridge
 port access vlan 10
 combo enable fiber
#
interface GigabitEthernet1/0/6
 port link-mode bridge
```

```
 port access vlan 10
 combo enable fiber
#
interface GigabitEthernet1/0/7
 port link-mode bridge
 port access vlan 10
 combo enable fiber
#
interface GigabitEthernet1/0/8
 port link-mode bridge
 port access vlan 10
 combo enable fiber
#
interface GigabitEthernet1/0/9
 port link-mode bridge
 port access vlan 20
 combo enable fiber
#
interface GigabitEthernet1/0/10
 port link-mode bridge
 port access vlan 20
 combo enable fiber
#
interface GigabitEthernet1/0/11
 port link-mode bridge
 port access vlan 20
 combo enable fiber
#
interface GigabitEthernet1/0/12
 port link-mode bridge
 port access vlan 20
 combo enable fiber
#
interface GigabitEthernet1/0/13
 port link-mode bridge
 port access vlan 20
 combo enable fiber
#
interface GigabitEthernet1/0/14
 port link-mode bridge
 port access vlan 20
 combo enable fiber
#
interface GigabitEthernet1/0/15
 port link-mode bridge
```

```
 port access vlan 20
 combo enable fiber
#
interface GigabitEthernet1/0/16
 port link-mode bridge
 port access vlan 20
 combo enable fiber
#
interface GigabitEthernet1/0/17
 port link-mode bridge
 port access vlan 30
 combo enable fiber
#
interface GigabitEthernet1/0/18
 port link-mode bridge
 port access vlan 30
 combo enable fiber
#
interface GigabitEthernet1/0/19
 port link-mode bridge
 port access vlan 30
 combo enable fiber
#
interface GigabitEthernet1/0/20
 port link-mode bridge
 port access vlan 30
 combo enable fiber
#
interface GigabitEthernet1/0/21
 port link-mode bridge
 port access vlan 30
 combo enable fiber
#
interface GigabitEthernet1/0/22
 port link-mode bridge
 port access vlan 30
 combo enable fiber
#
interface GigabitEthernet1/0/23
 port link-mode bridge
 port link-type trunk
 port trunk permit vlan 1 10 20
 combo enable fiber
#
interface GigabitEthernet1/0/24
```

```
 port link-mode bridge
 combo enable fiber
#
interface GigabitEthernet1/0/25
 port link-mode bridge
 combo enable fiber
#
interface GigabitEthernet1/0/26
 port link-mode bridge
 combo enable fiber
#
interface GigabitEthernet1/0/27
 port link-mode bridge
 combo enable fiber
#
interface GigabitEthernet1/0/28
 port link-mode bridge
 combo enable fiber
#
interface GigabitEthernet1/0/29
 port link-mode bridge
 combo enable fiber
#
interface GigabitEthernet1/0/30
 port link-mode bridge
 combo enable fiber
#
interface GigabitEthernet1/0/31
 port link-mode bridge
 combo enable fiber
#
interface GigabitEthernet1/0/32
 port link-mode bridge
 combo enable fiber
#
interface GigabitEthernet1/0/33
 port link-mode bridge
 combo enable fiber
#
interface GigabitEthernet1/0/34
 port link-mode bridge
 combo enable fiber
#
interface GigabitEthernet1/0/35
 port link-mode bridge
```

```
 combo enable fiber
#
interface GigabitEthernet1/0/36
 port link-mode bridge
 combo enable fiber
#
interface GigabitEthernet1/0/37
 port link-mode bridge
 combo enable fiber
#
interface GigabitEthernet1/0/38
 port link-mode bridge
 combo enable fiber
#
interface GigabitEthernet1/0/39
 port link-mode bridge
 combo enable fiber
#
interface GigabitEthernet1/0/40
 port link-mode bridge
 combo enable fiber
#
interface GigabitEthernet1/0/41
 port link-mode bridge
 combo enable fiber
#
interface GigabitEthernet1/0/42
 port link-mode bridge
 combo enable fiber
#
interface GigabitEthernet1/0/43
 port link-mode bridge
 combo enable fiber
#
interface GigabitEthernet1/0/44
 port link-mode bridge
 combo enable fiber
#
interface GigabitEthernet1/0/45
 port link-mode bridge
 combo enable fiber
#
interface GigabitEthernet1/0/46
 port link-mode bridge
 combo enable fiber
```

```
#
interface GigabitEthernet1/0/47
 port link-mode bridge
 combo enable fiber
#
interface GigabitEthernet1/0/48
 port link-mode bridge
 combo enable fiber
#
interface M-GigabitEthernet0/0/0
#
interface Ten-GigabitEthernet1/0/49
 port link-mode bridge
 combo enable fiber
#
interface Ten-GigabitEthernet1/0/50
 port link-mode bridge
 combo enable fiber
#
interface Ten-GigabitEthernet1/0/51
 port link-mode bridge
 combo enable fiber
#
interface Ten-GigabitEthernet1/0/52
 port link-mode bridge
 combo enable fiber
#
 scheduler logfile size 16
#
line class aux
 user-role network-operator
#
line class console
 user-role network-admin
#
line class tty
 user-role network-operator
#
line class vty
 user-role network-operator
#
line aux 0
 user-role network-operator
#
line con 0
```

```
 user-role network-admin
#
line vty 0 63
 user-role network-operator
#
radius scheme system
 user-name-format without-domain
#
domain name system
#
 domain default enable system
#
role name level-0
 description Predefined level-0 role
#
role name level-1
 description Predefined level-1 role
#
role name level-2
 description Predefined level-2 role
#
role name level-3
 description Predefined level-3 role
#
role name level-4
 description Predefined level-4 role
#
role name level-5
 description Predefined level-5 role
#
role name level-6
 description Predefined level-6 role
#
role name level-7
 description Predefined level-7 role
#
role name level-8
 description Predefined level-8 role
#
role name level-9
 description Predefined level-9 role
#
role name level-10
 description Predefined level-10 role
#
```

```
role name level-11
 description Predefined level-11 role
#
role name level-12
 description Predefined level-12 role
#
role name level-13
 description Predefined level-13 role
#
role name level-14
 description Predefined level-14 role
#
user-group system
#
return
```

② 根据以上配置信息，整理出 S1、S2、S3 的 VLAN 信息和相关接口信息，并将这些信息填入表 5-2。

表 5-2　VLAN 信息和相关接口信息

交换机	已有 VLAN ID	相关接口信息
S1	VLAN 1	默认，G1/0/23 和 G1/0/24 接口是 trunk 接口
	VLAN 10	虚拟接口地址（SVI）：192.168.10.254/24
	VLAN 20	虚拟接口地址（SVI）：192.168.20.254/24
	VLAN 30	虚拟接口地址（SVI）：192.168.30.254/24
S2	VLAN 1	默认，G1/0/24 接口是 trunk 接口
	VLAN 10	归属接口有 G1/0/1～G1/0/8
	VLAN 20	归属接口有 G1/0/9～G1/0/17
	VLAN 30	归属接口有 G1/0/18～G1/0/23
S3	VLAN 1	默认，G1/0/23 接口是 trunk 接口
	VLAN 10	归属接口有 G1/0/1～G1/0/8
	VLAN 20	归属接口有 G1/0/9～G1/0/16
	VLAN 30	归属接口有 G1/0/17～G1/0/22

2）根据相关知识点及表 5-2，确定通信故障点

① 分别查看新增终端主机 PC5 和 PC6 的 IP 地址与规划地址是否一致。

检查结果为不存在故障点。

② 分析 S2、S3 的配置信息与规划的是否相符。

从表 5-2 中可以看出，S2 的 G1/0/17 接口被划分给了 VLAN 20，这里存在一个故障点，应该将该接口划分给 VLAN 30。

③分析 VLAN 跨交换机通信的中继链路配置问题，观察是否允许新增 VLAN 通过。

从表 5-2 中可以看出，在交换机互联链路上，S1、S2、S3 互联接口的 trunk 配置中都只允许 VLAN 1、VLAN 10、VLAN 20 通过，没有新增的 VLAN 30 。因此，这里也存在

一个故障点。

3）修改配置，保存配置信息并进行测试

① 修改配置，并保存配置信息。

首先，在 S2 上修改 G1/0/17 接口所属 VLAN，命令如下：

```
[S2]inter g 1/0/17
[S2-GigabitEthernet1/0/17]port access vlan 30
```

然后，将 S1、S2、S3 这 3 台交换机的 trunk 链路设置为允许 VLAN 30 通过，命令如下：

```
[S1]inter range GigabitEthernet 1/0/23 to GigabitEthernet 1/0/24
[S1-if-range]port trunk permit vlan all

[S2]inter g 1/0/24
[S2-GigabitEthernet1/0/24]port trunk permit vlan all

[S3]inter g 1/0/23
[S3-GigabitEthernet1/0/23]port trunk permit vlan 30
```

最后，保存各设备的配置信息。

② 在新配置环境下进行测试，结果如表 5-3 所示。

表 5-3　新配置环境下的测试结果

测试序号	交换机 S2	交换机 S3	测试方法	预期测试结果	实际测试结果	是否发生故障
1	PC1、PC5		ping 命令	成功	成功	否
2	PC2、PC5		ping 命令	成功	成功	否
3		PC3、PC6	ping 命令	成功	成功	否
4		PC4、PC6	ping 命令	成功	成功	否
5	PC5	PC6	ping 命令	成功	成功	否

③ 整理新的配置文档。在故障排除后，保存所有交换机的配置信息，并更新书面的记录材料，确保书面文档与实际配置保持一致，以确保下次配置正常使用。

5.3　相关知识准备

知识准备

为了能够深入地分析故障点，读者应了解三层交换机的相关知识。

三层交换机就是具有部分路由器功能的交换机，工作在 OSI（Open System Interconnection，开放系统互连）参考模型的网络层，能够做到"一次路由，多次转发"。

SVI（Switch Virtual Interface，交换机虚拟接口）是连接 VLAN 的 IP 地址接口，一个 SVI 只能与一个 VLAN 相连。SVI 有如下两种类型。

① 主机管理接口：网络管理员可以利用该接口管理交换机。

② 网关接口：用于三层交换机不同 VLAN 间的路由。首先使用 Interface VLAN 接口配置命令来创建 SVI，然后为其配置 IP 地址即可开启路由功能。

三层交换机最重要的作用是加快大型局域网内部的数据交换。对于数据包转发等有规律性过程的功能，其由硬件高速实现；对于路由信息更新、路由表维护、路由计算、路由确定等功能，其由软件实现。三层交换技术=二层交换+三层转发技术。传统交换技术是在OSI参考模型的第二层——数据链路层进行操作的。三层交换技术在OSI参考模型的第三层——网络层实现了数据包的高速转发，既可实现网络路由功能，又可根据不同网络状况使网络性能达到最优。三层交换机具备如下特点。

1. 高可扩充性

三层交换机在连接多个子网时，子网只是与第三层交换模块建立逻辑连接，而不像传统外接路由器那样需要增加接口，因此能够节约网络投资。

2. 高性价比

三层交换机具有连接大型网络的能力，功能上基本可以取代某些传统路由器，但是价格却接近二层交换机。

3. 内置安全机制

三层交换机与普通路由器一样，具有访问列表的功能，可以实现不同VLAN间的单向或双向通信。如果在访问列表中进行设置，那么三层交换机还可以限制用户访问特定的IP地址。

5.4 项目小结

本项目介绍了对VLAN间路由故障的排除，主要思路是先解决同一VLAN内的通信故障，再对VLAN间路由故障进行分析，涉及的知识点主要包括中继链路的设置和接口VLAN的归属。在实际环境中，编者遇到过其他原因导致的VLAN间路由故障，如核心网络三层交换机上没有启用路由功能，以及当交换机转发某个VLAN信息时，此交换机上没创建这个VLAN。另外，本项目中最基本和最关键的一个操作就是IP地址的配置，往往最简单的地方也是最容易忽略的地方。

素质拓展：炽热青春 开拓创新

"艰苦创业、奋发图强、无私奉献、开拓创新"是大陈岛垦荒精神的主要内涵。

大陈岛第一个5G电话于2019年5月14日拨通。此后，远程医疗、智慧教学、智慧旅游等应用场景应运而生。2021年8月，随着海底光缆的前期敷设工作全面完成，大陈岛5G信号全覆盖计划迈出重要一步。"互联网+数字化"模式下的"三医联动"便民服务点，就医、买药和报销最短只需20min。线上"垦荒邮局"畅通了"'海岛好货'出岛、'城里优品'进岛"的通道。开拓创新推动着大陈岛的发展。

增值服务

在类似案例中，VLAN 一般分为用户 VLAN 和管理 VLAN。从安全角度和工程角度出发，人们往往把用户 VLAN 和管理 VLAN 规划为不同的网段。因此，当在交换机上配置 SVI 地址作为 PC 网关时，不能忽略管理 VLAN 的数据通信。二层交换机作为接入设备，其默认网关是初级网络管理员容易忽视的一个细节，这将影响接入设备的远程维护实现。通过对上述案例进行故障诊断与排查，建议在本案例中增加交换机 S2 与 S3 的默认网关，以解决日后远程运维时可能出现的隐藏故障问题。

5.5 课后实训

项目内容：某市统计局的局域网主要由核心交换机 S1 及接入交换机 S2 和 S3 构成，并且核心交换机 S1 连接了接入交换机 S2 和 S3。目前存在的业务主要有 VLAN 10 和 VLAN 20，它们使用的网段地址分别为 192.168.100.0/24 和 192.168.200.0/24（网关设置在核心交换机 S1 上，IP 地址分别为 192.168.100.254 和 192.168.200.254）。最近根据业务需求，需要在局域网中增加业务 VLAN 30，并将接入交换机 S2 的 F0/17～F0/24 接口与接入交换机 S3 的 F0/17～F0/24 接口划入 VLAN 30，将 VLAN 30 的网关设置在核心交换机 S1 上。使用网段 192.168.30.0/24 为 VLAN 30 分配地址，并且网关地址为 192.168.30.254。要求 VLAN 30 内部设备之间能够互联互通，VLAN 30 的业务与 VLAN 10 和 VLAN 20 的业务也能互联互通。

在完成网络配置后进行测试时，出现了故障现象。请按如下操作步骤排除故障。

① 根据要求检查故障现象。
② 根据故障现象收集故障信息。
③ 利用结构化故障排除方法完成故障定位。
④ 修改故障配置并说明故障原因。
⑤ 更新配置文档。

项目 6

VRRP 应用的故障排除

内容介绍

某公司为 Internet 用户提供 WWW 服务及 E-mail 服务，为了防止因服务器网关失效而造成业务中断，服务器侧部署了双网关，并使用了 VRRP，以避免在网关处发生单点故障。

任务安排

任务1　了解单点故障容错协议 VRRP 的应用
任务2　进行网络业务更新过程中的故障分析与排除

学习目标

◆ 了解 VRRP 常见故障的产生原因
◆ 掌握故障排除的思路
◆ 学会结构化故障排除方法
◆ 学会 VRRP 相关故障排除及文档更新的方法

素质目标

建立良好的客户关系，提高客户满意度，与客户进行有效沟通，解决紧急事件。

6.1 VRRP 配置分析与实施

发现故障

某公司为 Internet 用户提供 WWW 服务及 E-mail 服务,为了防止因服务器网关失效而造成业务中断,服务器侧部署了双网关,并使用了 VRRP,以避免在网关处发生单点故障,具体网络拓扑结构如图 6-1[①]所示。网络管理员李工顺利地实现了公司需求,并且网络稳定运行了一段时间。最近,由于新业务拓展,公司网络管理员在交换机 S2 上做了部分操作,导致提供服务的 WWW 服务器与 E-mail 服务器发生网络故障。

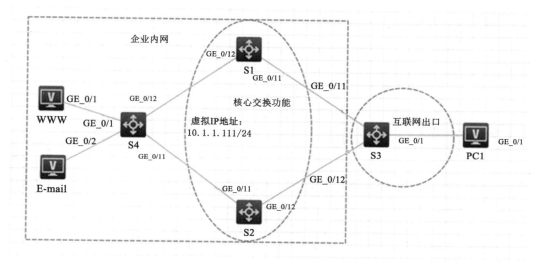

图 6-1 网络拓扑结构

李工上门服务,针对该公司的网络故障报告,在内网从 WWW 服务器和 E-mail 服务器上分别 ping 核心交换机,测试结果如表 6-1 所示。

表 6-1 故障测试结果

序号	终端	核心交换机	测试方法	预期测试结果	实际测试测试	是否发生故障
1	WWW 服务器	S1	ping 命令	成功	失败	是
2	WWW 服务器	S2	ping 命令	成功	失败	是
3	E-mail 服务器	S1	ping 命令	成功	失败	是
4	E-mail 服务器	S2	ping 命令	成功	失败	是

表 6-1 的测试结果表明服务器网关没有联通。这是什么原因造成的故障现象呢?随后,李工根据结构化故障排除方法,并结合 VRRP 的技术原理和实际操作步骤,制订了以下操作计划。

① 使用 disp vrrp verbose 命令查看 VRRP 运行状态。
② 使用 ping 命令测试服务器与核心交换机物理接口之间的连通性。

---
① 本书使用 GE_表示 GigabitEthernet。

③ 查看当前运行的核心交换机的配置。
④ 修改错误配置，并保存配置信息。
⑤ 在新配置环境下进行测试。
⑥ 整理新的配置文档。

6.2 VRRP 配置故障分析与排除

1. 故障分析方法

根据结构化故障排除思路，严格执行故障排除的操作步骤。首先，确定故障现象并进行详细记录；其次，收集设备信息，本项目主要收集交换机的配置信息、IP 地址、接口信息、VRRP 运行状态等；再次，在收集完信息后，结合 VRRP 的技术原理进行综合分析，确定并罗列可能存在的故障点；最后，针对故障点分析出最有可能的故障原因，并对这个原因进行故障排除。

2. 分析故障点

从 VRRP 的实现及技术原理进行分析，可能存在以下故障点。
① VRRP 停用或存在错误的 IP 地址。
② VRRP 中的 Master 与 Backup 状态配置错误。
③ 有错误的优先级。
④ VRID 错误。

1）查看 VRRP 运行状态
（1）任务目标。
①查看核心交换机 S1 的配置，并确定 VRRP 运行状态信息。
②查看核心交换机 S2 的配置，并确定 VRRP 运行状态信息。
（2）任务所需设备。
① 一台装有超级终端软件或 Telnet 软件的计算机，同时确定访问所需的用户名和口令。
② 配置线缆。
③ 笔和纸，用于记录相关信息。
（3）具体实施。
① 使用超级终端软件或 Telnet 软件连接交换机，并使用 disp vrrp verbose 命令分别查看两台核心交换机 S1 和 S2 的 VRRP 运行状态信息。
查看核心交换机 S1 的 VRRP 运行状态信息：

```
[H3C]disp vrrp verbose
IPv4 virtual router information:
 Running mode    : Standard
```

```
Total number of virtual routers : 1
  Interface Vlan-interface20
    VRID              : 1              Adver timer   : 100 centiseconds
    Admin status      : Up             State         : Master
    Config pri        : 110            Running pri   : 110
    Preempt mode      : Yes            Delay time    : 500 centiseconds
    Auth type         : None
    Virtual IP        : 10.1.0.111
    Virtual MAC       : 0000-5e00-0101
    Master IP         : 10.1.0.1
```

查看核心交换机 S2 的 VRRP 运行状态信息：

```
[H3C]disp vrrp verbose
IPv4 virtual router information:
 Running mode    : Standard
 Total number of virtual routers : 1
  Interface Vlan-interface20
    VRID              : 1              Adver timer   : 100 centiseconds
    Admin status      : Up             State         : Initialize
    Config pri        : 100            Running pri   : 100
    Preempt mode      : Yes            Delay time    : 500 centiseconds
    Auth type         : None
    Virtual IP        : 10.1.1.111
    Master IP         : 0.0.0.0
```

② 根据以上信息，整理出核心交换机 S1 和 S2 的 VRRP 运行状态信息，并将这些信息填入表 6-2。

表 6-2　核心交换机 S1 和 S2 的 VRRP 运行状态信息

核心交换机	接口	VRRP 备份组	优先级	状态	虚拟 IP 地址
S1	Vlan-interface20	1	110	Master	10.1.0.111
S2	Vlan-interface20	1	100	Initialize	10.1.1.111

根据表 6-2 中的信息，可看出两台核心交换机的 VRRP 备份组同为 1，一个状态为 Master，另一个状态为 Initialize，而正常状态下应该一个为 Master，另一个为 Backup；两台核心交换机所配置的虚拟 IP 地址也不相同，而正常状态下应该相同。因此，需要使用其他方式排查导致这种状态的原因。

2）使用 ping 命令测试网络连通性，核查各设备接口 IP 地址配置情况

复核公司的网络故障报告，测试网络连通性，在报告属实的情况下，核查 IP 地址配置。

① 查看 WWW 服务器和 E-mail 服务器的 IP 地址配置，如图 6-2 所示。

图 6-2 WWW 服务器和 E-mail 服务器的 IP 地址配置

② 查看核心交换机 S1 的 IP 地址配置：

```
[S1]disp ip inter bri
*down: administratively down
(s): spoofing  (l): loopback
Interface          Physical      Protocol     IP Address        Description
MGE0/0/0           down          down         --                --
Vlan20             up            up           10.1.0.1          --
Vlan30             down          down         10.1.3.1          --
```

③ 查看核心交换机 S2 的 IP 地址配置：

```
[S2]disp ip inter bri
*down: administratively down
(s): spoofing  (l): loopback
Interface          Physical      Protocol     IP Address        Description
MGE0/0/0           down          down         --                --
Vlan20             *down         down         10.1.1.2          --
Vlan30             down          down         10.1.5.1          --
```

④ 查看交换机 S3 的 IP 地址配置[①]：

```
[S3] disp ip inter bri
*down: administratively down
(s): spoofing  (l): loopback
Interface          Physical      Protocol     IP Address        Description
GE1/0/11           *down         down         10.1.3.2          --
GE1/0/12           *down         down         10.1.4.2          --
MGE0/0/0           down          down         --                --
Vlan20             down          down         10.1.2.2          --
```

---

① 本书使用 GE 表示 GigabitEthernet。

⑤ 将查看的结果与原规划 IP 地址进行对比，并将相关信息填入表 6-3。

表 6-3 设备/接口的 IP 地址信息

设备/接口	运行地址	规划地址	测试结果	备注
WWW 服务器	10.1.1.3	10.1.1.3	一致	网关正确
E-mail 服务器	10.1.1.4	10.1.1.4	一致	网关正确
S1-VLAN 20	10.1.0.1	10.1.1.1	不一致	up
S1-VLAN 30	10.1.3.1	10.1.3.1	一致	down
S2-VLAN 20	10.1.1.2	10.1.1.2	一致	down
S2-VLAN 30	10.1.5.1	10.1.4.1	不一致	down
S3-VLAN 20	10.1.2.2	10.1.2.2	一致	down
S3-G1/0/11	10.1.3.2	10.1.3.2	一致	down
S3-G1/0/12	10.1.4.2	10.1.4.2	一致	down

从表 6-3 中可以看出，网络管理员在配置设备时要么关闭了接口，要么没有按规划配置 IP 地址，或者在配置过程中发生了误配。因此，接下来需要手动开启接口或修改 IP 地址。

3）修改错误配置，并保存配置信息

① 开启交换机 S3 的 G1/0/11 接口和 G1/0/12 接口：

```
[S3]int GE1/0/11
[S3-GigabitEthernet1/0/11]undo shut
[S3-GigabitEthernet1/0/11]int GE1/0/12
[S3-GigabitEthernet1/0/12]undo shut
```

结果如下：

```
[S3]disp ip inter bri
*down: administratively down
(s): spoofing  (l): loopback
Interface         Physical      Protocol      IP Address        Description
GE1/0/11          up            up            10.1.3.2          --
GE1/0/12          up            up            10.1.4.2          --
MGE0/0/0          down          down          --                --
Vlan20            up            up            10.1.2.2          --
```

② 修改核心交换机 S2 的 VLAN 30 的地址，并且启动 VLAN 20 和 VLAN 30：

```
[S2]int VLAN 30
[S2-Vlan-interface30]ip add 10.1.4.1 24
[S2]inter vlan 20
[S2-Vlan-interface20]undo shut
```

结果如下：

```
[S2]disp ip inter bri
*down: administratively down
(s): spoofing  (l): loopback
Interface         Physical      Protocol      IP Address        Description
MGE0/0/0          down          down          --                --
Vlan20            up            up            10.1.1.2          --
```

```
Vlan30                       up            up             10.1.4.1           --
[S2]disp vrrp verbose
IPv4 virtual router information:
 Running mode    : Standard
 Total number of virtual routers : 1
   Interface Vlan-interface20
     VRID             : 1              Adver timer    : 100 centiseconds
     Admin status     : Up             State          : Backup
     Config pri       : 100            Running pri    : 100
     Preempt mode     : Yes            Delay time     : 500 centiseconds
     Become master    : 2780 millisecond left
     Auth type        : None
     Virtual IP       : 10.1.1.111
     Master IP        : 10.1.1.1
```

③ 将核心交换机 S1 的 G1/0/11 接口加入 VLAN 30，并且修改 VLAN 20 和 Virtual-IP 地址：

```
[S1]vlan 30
[S1-vlan30]port GE1/0/11
[S1]inter vlan 20
[S1-Vlan-interface20]ip address 10.1.1.1 24
[S1]inter vlan 20
[S1-Vlan-interface20]vrrp vrid 1 virtual-ip 10.1.1.111
[S1-Vlan-interface20]undo vrrp vrid 1 virtual-ip 10.1.0.111
```

结果如下：

```
[S1]disp vrrp verbose
IPv4 virtual router information:
 Running mode    : Standard
 Total number of virtual routers : 1
   Interface Vlan-interface20
     VRID             : 1              Adver timer    : 100 centiseconds
     Admin status     : Up             State          : Master
     Config pri       : 110            Running pri    : 110
     Preempt mode     : Yes            Delay time     : 500 centiseconds
     Auth type        : None
     Virtual IP       : 10.1.1.111
     Virtual MAC      : 0000-5e00-0101
     Master IP        : 10.1.1.1
[S1]disp ip inter bri
*down: administratively down
(s): spoofing (l): loopback
Interface              Physical      Protocol       IP Address         Description
MGE0/0/0               down          down           --                 --
Vlan20                 up            up             10.1.1.1           --
Vlan30                 up            up             10.1.3.1           --
```

4）在新配置环境下进行测试

① 测试网络连通性，并将结果填入表 6-4。

表 6-4　连通性测试结果（1）

测试序号	设备	测试方法	预期测试结果	实际测试结果	是否发生故障
1	WWW 服务器	ping 命令	成功	成功	否
2	E-mail 服务器	ping 命令	成功	成功	否

② 关闭核心交换机 S1 的 G1/0/12 接口，测试 WWW 服务器和 E-mail 服务器与 PC1 之间的连通性，并将结果填入表 6-5。

表 6-5　连通性测试结果（2）

测试序号	设备	测试方法	预期测试结果	实际测试结果
1	WWW 服务器	ping 命令	成功	成功
2	E-mail 服务器	ping 命令	成功	成功

③ 关闭核心交换机 S2 的 G1/0/11 接口，开启核心交换机 S1 的 G1/0/12 接口，测试 WWW 服务器和 E-mail 服务器与 PC1 之间的连通性，并将结果填入表 6-6。

表 6-6　连通性测试结果（3）

测试序号	设备	测试方法	预期测试结果	实际测试结果
1	WWW 服务器	ping 命令	成功	成功
2	E-mail 服务器	ping 命令	成功	成功

5）整理新的配置文档

在故障排除后，保存所有交换机的配置信息，并更新书面记录材料，确保书面文档和实际配置保持一致，以确保下次配置正常使用。

6.3　相关知识准备

为了能够深入地分析故障点，读者应了解 VRRP 的相关知识。

1）什么是 VRRP

VRRP（Virtual Router Redundancy Protocol，虚拟路由冗余协议）是防止路由器出现单点故障的一种容错协议。VRRP 保证当主机的下一跳路由器出现故障时，由另一台路由器来代替出现故障的路由器进行工作，从而保持网络通信的连续性和可靠性。

VRRP 将局域网内的一组路由器划分在一起，形成一个 VRRP 备份组，其在功能上相当于一台虚拟路由器，使用虚拟路由器号进行标识。

虚拟路由器有自己的虚拟 IP 地址和虚拟 MAC 地址，其外在表现形式与物理路由器完全一样。局域网内的主机将虚拟路由器的 IP 地址设置为默认网关，通过虚拟路由器与外部网络进行通信。

虚拟路由器工作在物理路由器上，由多个物理路由器组成，包括一台 Master 路由器和多台 Backup 路由器。Master 路由器在正常工作时，局域网内的主机通过 Master 路由器与外界进行通信。当 Master 路由器出现故障时，Backup 路由器中的一台设备将成为新的 Master 路由器，接替转发报文任务，如图 6-3 所示。

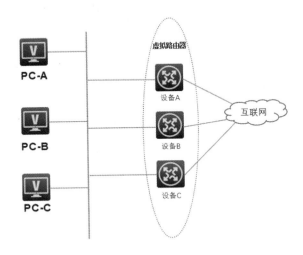

图 6-3　虚拟路由器示意图

与 VRRP 相关的常用术语如下。

① 虚拟路由器：由一台 Master 路由器和多台 Backup 路由器组成。局域网内的主机会将虚拟路由器的 IP 地址设置为默认网关。

② VRID：虚拟路由器的标识。一组具有相同 VRID 的路由器可以构成一个虚拟路由器。

③ Master 路由器：虚拟路由器中承担报文转发任务的路由器。

④ Backup 路由器：当 Master 路由器出现故障时，能够接替 Master 路由器工作的路由器。

⑤ 虚拟 IP 地址：虚拟路由器的 IP 地址。一个虚拟路由器可以拥有一个或多个 IP 地址。

⑥ 虚拟 MAC 地址：一个虚拟路由器拥有一个虚拟 MAC 地址。虚拟 MAC 地址的格式为 00-00-5E-00-01-{VRID}。在通常情况下，虚拟路由器在回应 ARP 请求时使用的是虚拟 MAC 地址，只有当虚拟路由器做特殊配置时，才回应真实接口的 MAC 地址。

⑦ 优先级：VRRP 根据优先级来确定虚拟路由器中每台路由器的地位。

⑧ 非抢占方式：如果 Backup 路由器工作在非抢占方式下，则只要 Master 路由器没有出现故障，Backup 路由器即使被配置了更高的优先级也不会成为 Master 路由器。

⑨ 抢占方式：如果 Backup 路由器工作在抢占方式下，那么当它收到 VRRP 报文后，会将自己的优先级与 VRRP 报文中的优先级进行比较。若自己的优先级比当前 Master 路由器的优先级高，则会主动抢占成为 Master 路由器，否则将保持 Backup 状态。

2）VRRP 的优点

① 简化网络管理。在具有多播或广播能力的局域网（如以太网）中，利用 VRRP 能在

某台设备出现故障时仍然提供高可靠的默认链路,有效避免在单一链路发生故障后网络中断的问题。此外,使用 VRRP 不需要修改动态路由协议、路由发现协议等配置信息,也不需要修改主机的默认网关配置。

② 适应性强。VRRP 报文被封装在 IP 报文中,支持各种上层协议。

③ 网络开销小。VRRP 只定义了一种报文——VRRP 通告报文,并且只有处于 Master 状态的路由器才可以发送 VRRP 报文。

3)VRRP 的工作过程

① 虚拟路由器根据优先级选举出 Master 路由器。Master 路由器通过发送免费 ARP 报文,将自己的虚拟 MAC 地址通知给与它相连的设备或主机,从而承担报文转发任务。

② Master 路由器会定期发送 VRRP 报文,以公布自己的配置信息(优先级等)和工作状态。

③ 如果 Master 路由器出现故障,那么虚拟路由器中的 Backup 路由器将根据优先级重新选举 Master 路由器。

④ 当 Backup 路由器的优先级高于 Master 路由器的优先级时,由 Backup 路由器的工作方式(抢占方式和非抢占方式)决定是否重新选举 Master 路由器。

6.4 项目小结

本项目主要针对 VRRP 应用进行故障排除,介绍了当局域网出现故障时应该采用的故障排除方法和步骤。在 VRRP 工作过程中,Master 路由器会定期发送 VRRP 报文,并在 VRRP 备份组中公布自己的配置信息和工作状态。Backup 路由器通过接收到的 Master 路由器发送的 VRRP 报文来判断 Master 路由器是否正常工作,从而实现主、备设备的切换,确保网络链路信息传输的可靠性。

素质拓展:网络强国 国之大者

建设网络强国是网信事业的"国之大者"。信息技术浪潮气象万千,数字经济发展生机勃勃。

网信事业代表着新的生产力、新的发展方向。从过去的建设农业强国、工业强国、贸易强国、金融强国……到现在的建设网络强国,是适应时代发展之需,体现了人们对美好生活的向往和期盼。

从发布《中华人民共和国网络安全法》《中华人民共和国数据安全法》《中华人民共和国个人信息保护法》《关键信息基础设施安全保护条例》《国家网络空间安全战略》等网络安全法律、法规,到印发《关于加强网络安全学科建设和人才培养的意见》《关于加强国家网络安全标准化工作的若干意见》等政策文件,这一系列举措凝聚了改革的共识、发展的共识、法治的共识、核心价值观的共识,突显了依法推进网络强国战略的实践要求,突显了让互联网更好造福国家和人民的真挚情怀,突显了构建网络空间命运共同体的执着坚毅。

从2021年世界互联网大会主题"迈向数字文明新时代——携手构建网络空间命运共同体"到2022年世界互联网大会主题"共建网络世界 共创数字未来——携手构建网络空间命运共同体",这是为构建网络空间命运共同体持续"添柴加火",为全球互联网共享共治创造新的重要契机,让中国以更自信、更有力、更坚定、更迅疾的步伐,向网络强国昂首迈进,为携手构建网络空间命运共同体展现中国智慧、中国担当、中国力量。

增值服务

为了满足用户对网络高可靠性的要求,某公司原有网络采取了双机热备份路由器,配置了VRRP。然而,VRRP中主、备路由器的切换速度不够快,售后工程人员需要优化VRRP与BFD联动配置,实现快速切换和秒级故障感知;同时,为了防止主路由器的上行链路故障导致网络中断,需要采用BFD技术监视上行链路。

对本案例实现VRRP和BFD的联动,具体要求如下。

(1)配置动态BFD会话:在S1和S2上配置BFD会话,并使用动态会话监测S1与S2之间的链路。

(2)配置VRRP与BFD联动:在S2上配置VRRP与BFD联动,实现在发生链路故障时VRRP备份组的快速切换。

(3)配置静态BFD会话:在S1和S3上配置BFD会话,并使用静态会话监测S1与S3之间的链路。

(4)配置VRRP与BFD联动,监视上行链路:在S1上配置VRRP与BFD联动,实现当上行链路发生故障时触发VRRP备份组进行主、备路由器的切换。

6.5 课后实训

项目内容:某公司新成立的分公司A需要访问分公司B的网络,分公司A使用的网段地址为192.168.1.0/24,分公司B使用的网段地址为202.20.1.0/24。分公司A所在网络的出口处部署了两台设备,现要求使用VRRP的备份组功能,将这两台设备组成一台虚拟路由器,作为分公司A的默认网关。目前已经完成分公司A交换机的接口配置及Vlan-interface接口基本配置。请根据故障现象,完成以下任务。

① 根据要求检查故障现象。
② 根据故障现象收集故障信息。
③ 利用结构化故障排除方法完成故障定位。
④ 修改故障配置并说明故障原因。
⑤ 更新配置文档。

项目 7

静态路由协议应用的故障排除

内容介绍

某公司刚刚成立不久，没有自己的内部局域网，现在急需建设一个局域网来满足办公需求。考虑到公司刚成立，资金有限，业务规模较小，因此规划采用静态路由实现内部网络的互联互通，以便管理和维护。

任务安排

任务1　按规划新建局域网，并进行网络配置和测试
任务2　进行网络配置过程中的故障分析与排除

学习目标

◇ 了解网络管理的概念
◇ 了解静态路由的配置方法
◇ 掌握故障排除的思路
◇ 学会结构化故障排除方法
◇ 学会如何排除由静态路由配置问题引起的故障及文档更新的方法

素质目标

注重服务意识，责任心强，性格开朗，能够勤奋工作，具有一定的抗压能力，树立正确的职业观，敢于承担责任。

7.1 静态路由配置分析与实施

发现故障

网络管理员王工在接到该公司网络建设任务后,分析了该公司的业务和需求,制订了以下操作计划。

① 规划网络拓扑结构。
② 规划 IP 地址。
③ 规划路由器、交换机所连接的接口。
④ 对路由器、交换机进行配置。
⑤ 进行终端互联互通测试。

1)规划网络拓扑结构

网络管理员王工在分析了该公司业务和需求后,规划了网络拓扑结构,如图 7-1 所示。

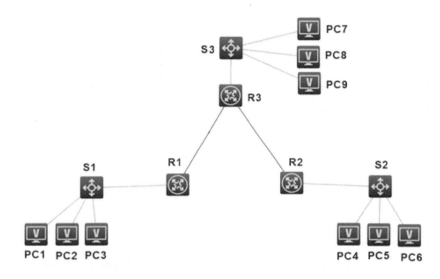

图 7-1 网络拓扑结构

从图 7-1 中可以看出,PC1、PC2、PC3 在同一个网段,PC4、PC5、PC6 在同一个网段,PC7、PC8、PC9 在同一个网段。路由器 R1 与路由器 R3 之间的互联接口是 S1/0[①],路由器 R2 与路由器 R3 之间的互联接口是 S2/0;路由器 R1 的 G0/1 接口与交换机 S1 的 G0/1 接口互连,路由器 R2 的 G0/1 接口与交换机 S2 的 G0/1 接口互连,路由器 R3 的 G0/0 接口与交换机 S3 的 G0/1 接口互连;PC1、PC2、PC3 分别与交换机 S1 的 G0/2、G0/3、G0/4 接口互连,PC4、PC5、PC6 分别与交换机 S2 的 G0/2、G0/3、G0/4 接口互连,PC7、PC8、PC9 分别与交换机 S3 的 G0/2、G0/3、G0/4 接口互连。

2)规划 IP 地址

参照图 7-1,使用子网划分方法进行 IP 地址规划,如表 7-1 所示。

---

① 本书使用 S 表示 Serial。

表 7-1　各接口 IP 地址规划

路由器 R1 与路由器 R3 的互连地址	路由器 R2 与路由器 R3 的互连地址	路由器 R1 上的网关地址、业务地址	路由器 R2 上的网关地址、业务地址	路由器 R3 上的网关地址、业务地址
172.16.1.2/30	172.16.2.2/30	网关地址：192.168.1.1/24	网关地址：192.168.2.1/24	网关地址：192.168.3.1/24
172.16.1.1/30	172.16.2.1/30	业务地址：192.168.1.0/24	业务地址：192.168.2.0/24	业务地址：192.168.3.0/24

3）规划设备连接接口，并对路由器、交换机进行配置

配置结果如下。查看路由器 R1 的配置：

```
[R1]disp cu
#
 version 7.1.075, Alpha 7571
#
 sysname R1
#
 system-working-mode standard
 xbar load-single
 password-recovery enable
 lpu-type f-series
#
vlan 1
#
interface Serial1/0
 ip address 172.16.1.2 255.255.255.252
#
interface Serial2/0
#
interface Serial3/0
#
interface Serial4/0
#
interface NULL0
#
interface GigabitEthernet0/0
 port link-mode route
 combo enable copper
#
interface GigabitEthernet0/1
 port link-mode route
 combo enable copper
 ip address 192.168.1.1 255.255.255.0
#
interface GigabitEthernet0/2
```

```
 port link-mode route
 combo enable copper
#
interface GigabitEthernet5/0
 port link-mode route
 combo enable copper
#
interface GigabitEthernet5/1
 port link-mode route
 combo enable copper
#
interface GigabitEthernet6/0
 port link-mode route
 combo enable copper
#
interface GigabitEthernet6/1
 port link-mode route
 combo enable copper
#
 scheduler logfile size 16
#
line class aux
 user-role network-operator
#
line class console
 user-role network-admin
#
line class tty
 user-role network-operator
#
line class vty
 user-role network-operator
#
line aux 0
 user-role network-operator
#
line con 0
 user-role network-admin
#
line vty 0 63
 user-role network-operator
#
ip route-static 172.16.2.0 30 Serial1/0
ip route-static 192.168.2.0 24 Serial1/0
#
```

```
domain name system
#
 domain default enable system
#
role name level-0
 description Predefined level-0 role
#
role name level-1
 description Predefined level-1 role
#
role name level-2
 description Predefined level-2 role
#
role name level-3
 description Predefined level-3 role
#
role name level-4
 description Predefined level-4 role
#
role name level-5
 description Predefined level-5 role
#
role name level-6
 description Predefined level-6 role
#
role name level-7
 description Predefined level-7 role
#
role name level-8
 description Predefined level-8 role
#
role name level-9
 description Predefined level-9 role
#
role name level-10
 description Predefined level-10 role
#
role name level-11
 description Predefined level-11 role
#
role name level-12
 description Predefined level-12 role
#
role name level-13
 description Predefined level-13 role
```

```
#
role name level-14
 description Predefined level-14 role
#
user-group system
#
return
[R1]IP ROU
[R1]IP route-static 192.168.3.0 24 S1/0
[R1]disp cu
#
 version 7.1.075, Alpha 7571
#
 sysname R1
#
 system-working-mode standard
 xbar load-single
 password-recovery enable
 lpu-type f-series
#
vlan 1
#
interface Serial1/0
 ip address 172.16.1.2 255.255.255.252
#
interface Serial2/0
#
interface Serial3/0
#
interface Serial4/0
#
interface NULL0
#
interface GigabitEthernet0/0
 port link-mode route
 combo enable copper
#
interface GigabitEthernet0/1
 port link-mode route
 combo enable copper
 ip address 192.168.1.1 255.255.255.0
#
interface GigabitEthernet0/2
 port link-mode route
 combo enable copper
```

```
#
interface GigabitEthernet5/0
 port link-mode route
 combo enable copper
#
interface GigabitEthernet5/1
 port link-mode route
 combo enable copper
#
interface GigabitEthernet6/0
 port link-mode route
 combo enable copper
#
interface GigabitEthernet6/1
 port link-mode route
 combo enable copper
#
 scheduler logfile size 16
#
line class aux
 user-role network-operator
#
line class console
 user-role network-admin
#
line class tty
 user-role network-operator
#
line class vty
 user-role network-operator
#
line aux 0
 user-role network-operator
#
line con 0
 user-role network-admin
#
line vty 0 63
 user-role network-operator
#
 ip route-static 172.16.2.0 30 Serial1/0
 ip route-static 192.168.2.0 24 Serial1/0
 ip route-static 192.168.3.0 24 Serial1/0
#
domain name system
```

```
#
 domain default enable system
#
role name level-0
 description Predefined level-0 role
#
role name level-1
 description Predefined level-1 role
#
role name level-2
 description Predefined level-2 role
#
role name level-3
 description Predefined level-3 role
#
role name level-4
 description Predefined level-4 role
#
role name level-5
 description Predefined level-5 role
#
role name level-6
 description Predefined level-6 role
#
role name level-7
 description Predefined level-7 role
#
role name level-8
 description Predefined level-8 role
#
role name level-9
 description Predefined level-9 role
#
role name level-10
 description Predefined level-10 role
#
role name level-11
 description Predefined level-11 role
#
role name level-12
 description Predefined level-12 role
#
role name level-13
 description Predefined level-13 role
#
```

```
 role name level-14
  description Predefined level-14 role
 #
 user-group system
 #
return
```

查看路由器 R2 的配置：
```
[R2]disp cu
#
 version 7.1.075, Alpha 7571
#
 sysname R2
#
 system-working-mode standard
 xbar load-single
 password-recovery enable
 lpu-type f-series
#
vlan 1
#
interface Serial1/0
#
interface Serial2/0
 ip address 172.16.2.2 255.255.255.252
#
interface Serial3/0
#
interface Serial4/0
#
interface NULL0
#
interface GigabitEthernet0/0
 port link-mode route
 combo enable copper
 ip address 192.168.2.1 255.255.255.0
#
interface GigabitEthernet0/1
 port link-mode route
 combo enable copper
#
interface GigabitEthernet0/2
 port link-mode route
 combo enable copper
#
interface GigabitEthernet5/0
```

```
 port link-mode route
 combo enable copper
#
interface GigabitEthernet5/1
 port link-mode route
 combo enable copper
#
interface GigabitEthernet6/0
 port link-mode route
 combo enable copper
#
interface GigabitEthernet6/1
 port link-mode route
 combo enable copper
#
 scheduler logfile size 16
#
line class aux
 user-role network-operator
#
line class console
 user-role network-admin
#
line class tty
 user-role network-operator
#
line class vty
 user-role network-operator
#
line aux 0
 user-role network-operator
#
line con 0
 user-role network-admin
#
line vty 0 63
 user-role network-operator
#
 ip route-static 192.168.1.0 24 Serial2/0
 ip route-static 192.168.3.0 24 172.16.2.1
#
domain name system
#
 domain default enable system
#
```

```
role name level-0
 description Predefined level-0 role
#
role name level-1
 description Predefined level-1 role
#
role name level-2
 description Predefined level-2 role
#
role name level-3
 description Predefined level-3 role
#
role name level-4
 description Predefined level-4 role
#
role name level-5
 description Predefined level-5 role
#
role name level-6
 description Predefined level-6 role
#
role name level-7
 description Predefined level-7 role
#
role name level-8
 description Predefined level-8 role
#
role name level-9
 description Predefined level-9 role
#
role name level-10
 description Predefined level-10 role
#
role name level-11
 description Predefined level-11 role
#
role name level-12
 description Predefined level-12 role
#
role name level-13
 description Predefined level-13 role
#
role name level-14
 description Predefined level-14 role
#
```

```
user-group system
#
return
```
查看路由器 R3 的配置：
```
[r3]disp cu
#
 version 7.1.075, Alpha 7571
#
 sysname r3
#
 system-working-mode standard
 xbar load-single
 password-recovery enable
 lpu-type f-series
#
vlan 1
#
interface Serial1/0
 ip address 172.16.1.1 255.255.255.0
#
interface Serial2/0
 ip address 172.16.2.1 255.255.255.252
#
interface Serial3/0
#
interface Serial4/0
#
interface NULL0
#
interface GigabitEthernet0/0
 port link-mode route
 combo enable copper
 ip address 192.168.3.1 255.255.255.0
#
interface GigabitEthernet0/1
 port link-mode route
 combo enable copper
#
interface GigabitEthernet0/2
 port link-mode route
 combo enable copper
#
interface GigabitEthernet5/0
 port link-mode route
 combo enable copper
```

```
#
interface GigabitEthernet5/1
 port link-mode route
 combo enable copper
#
interface GigabitEthernet6/0
 port link-mode route
 combo enable copper
#
interface GigabitEthernet6/1
 port link-mode route
 combo enable copper
#
 scheduler logfile size 16
#
line class aux
 user-role network-operator
#
line class console
 user-role network-admin
#
line class tty
 user-role network-operator
#
line class vty
 user-role network-operator
#
line aux 0
 user-role network-operator
#
line con 0
 user-role network-admin
#
line vty 0 63
 user-role network-operator
#
 ip route-static 192.168.2.0 24 Serial2/0
#
domain name system
#
domain default enable system
#
role name level-0
 description Predefined level-0 role
#
```

```
role name level-1
 description Predefined level-1 role
#
role name level-2
 description Predefined level-2 role
#
role name level-3
 description Predefined level-3 role
#
role name level-4
 description Predefined level-4 role
#
role name level-5
 description Predefined level-5 role
#
role name level-6
 description Predefined level-6 role
#
role name level-7
 description Predefined level-7 role
#
role name level-8
 description Predefined level-8 role
#
role name level-9
 description Predefined level-9 role
#
role name level-10
 description Predefined level-10 role
#
role name level-11
 description Predefined level-11 role
#
role name level-12
 description Predefined level-12 role
#
role name level-13
 description Predefined level-13 role
#
role name level-14
 description Predefined level-14 role
#
user-group system
#
return
```

交换机的相关配置可参考本书配套素材，这里不进行细述。

4）进行终端互联互通测试

按照如图 7-1 所示的网络拓扑结构做好静态路由并分配好 IP 地址后，进行终端互联互通测试，结果如表 7-2 所示。

表 7-2　终端互联互通测试结果

测试序号	交换机 S1	交换机 S2	交换机 S3	测试方法	预期测试结果	实际测试结果	是否发生故障
1	PC1、PC2、PC3	PC4、PC5、PC6		ping 命令	成功	失败	是
2	PC1、PC2、PC3		PC7、PC8、PC9	ping 命令	成功	失败	是
3		PC4、PC5、PC6	PC7、PC8、PC9	ping 命令	成功	成功	否

由表 7-2 可知，有 2 项测试失败，因此本次操作没有成功完成项目目标。这是什么原因造成的故障现象呢？是规划设计的问题、操作的问题，还是静态路由概念没有理解清楚的问题？据此，要深入进行故障分析，以确定问题所在。

7.2　静态路由配置故障分析与排除

排除故障

1. 故障分析方法

根据结构化故障排除思路，故障排除需要遵循 OSI 参考模型的要求。这里采用自下而上的方法从物理层往上依次进行排除，同时使用与静态路由协议相关的技术进行综合分析。

从故障排除理论体系角度进行分析，故障点可能包括以下几方面：物理问题、损坏的电缆、损坏的接口、电源故障、设备问题、软件错误、性能问题、配置错误、路由协议故障。

2. 分析故障点

从静态路由的实现及技术原理进行分析，可能存在以下故障点。

① 下一跳地址指向错误或送出接口指向错误。
② 指向的目标网络地址、子网掩码错误。
③ 仅设置了单向路由，而没有设置双向路由。
④ 中途经过的中间设备没有源路由和目的路由。

由于本次故障主要通过模拟环境来实现，因此可以忽略物理问题及设备问题。若在实际环境下，则应该逐步进行故障排除。本次主要从静态路由协议的相关概念及静态路由协议的配置和操作方面进行故障排除。

完成本次故障排除任务所需的设备如下。

① 一台装有超级终端软件或 Telnet 软件的计算机，同时确定访问所需的用户名和口令。
② 配置线缆。
③ 笔和纸，用于记录相关信息。

1)查看 IP 地址配置(此处省略查看 PC 的 IP 地址配置)

查看路由器 R1 的 IP 地址配置:

```
[R1]disp ip inter bri
*down: administratively down
(s): spoofing  (l): loopback
Interface         Physical      Protocol I   P Address       Description
GE0/0             down          down         --              --
GE0/1             up            up           192.168.1.1     --
GE0/2             down          down         --              --
GE5/0             down          down         --              --
GE5/1             down          down         --              --
GE6/0             down          down         --              --
GE6/1             down          down         --              --
Ser1/0            up            up           172.16.1.2      --
Ser2/0            down          down         --              --
Ser3/0            down          down         --              --
Ser4/0            down          down         --              --
```

查看路由器 R2 的 IP 地址配置:

```
[R2]disp ip inter bri
*down: administratively down
(s): spoofing  (l): loopback
Interface         Physical      Protocol     IP Address      Description
GE0/0             up            up           192.168.2.1     --
GE0/1             down          down         --              --
GE0/2             down          down         --              --
GE5/0             down          down         --              --
GE5/1             down          down         --              --
GE6/0             down          down         --              --
GE6/1             down          down         --              --
Ser1/0            down          down         --              --
Ser2/0            up            up           172.16.2.2      --
Ser3/0            down          down         --              --
Ser4/0            down          down         --              --
```

查看路由器 R3 的 IP 地址配置:

```
[R3]disp ip inter bri
*down: administratively down
(s): spoofing  (l): loopback
Interface         Physical      Protocol     IP Address      Description
GE0/0             up            up           192.168.3.1     --
GE0/1             down          down         --              --
GE0/2             down          down         --              --
GE5/0             down          down         --              --
GE5/1             down          down         --              --
GE6/0             down          down         --              --
```

GE6/1	down	down	--	--
Ser1/0	up	up	172.16.1.1	--
Ser2/0	up	up	172.16.2.1	--
Ser3/0	down	down	--	--
Ser4/0	down	down	--	--

经过查看，上述路由器的 IP 地址配置都没有问题，PC 的 IP 地址也没问题，与规划时的一致，因此可以排除 IP 地址错误问题。

2）查看接口配置

经过查看，路由器 R1、路由器 R2、路由器 R3 的接口配置都是 up 状态，协议也是 up 状态，并且与规划时所用的接口一致，因此可以排除接口故障。

3）查看是否为静态路由下一跳地址指向错误或送出接口指向错误

查看路由器 R1 的路由表：

```
[R1]disp ip routing-table

Destinations  : 20       Routes : 20

Destination/Mask       Proto   Pre  Cost       nexthop          Interface
0.0.0.0/32             Direct  0    0          127.0.0.1        InLoop0
127.0.0.0/8            Direct  0    0          127.0.0.1        InLoop0
127.0.0.0/32           Direct  0    0          127.0.0.1        InLoop0
127.0.0.1/32           Direct  0    0          127.0.0.1        InLoop0
127.255.255.255/32     Direct  0    0          127.0.0.1        InLoop0
172.16.1.0/30          Direct  0    0          172.16.1.2       Ser1/0
172.16.1.0/32          Direct  0    0          172.16.1.2       Ser1/0
172.16.1.1/32          Direct  0    0          172.16.1.1       Ser1/0
172.16.1.2/32          Direct  0    0          127.0.0.1        InLoop0
172.16.1.3/32          Direct  0    0          172.16.1.2       Ser1/0
172.16.2.0/30          Static  60   0          0.0.0.0          Ser1/0
192.168.1.0/24         Direct  0    0          192.168.1.1      GE0/1
192.168.1.0/32         Direct  0    0          192.168.1.1      GE0/1
192.168.1.1/32         Direct  0    0          127.0.0.1        InLoop0
192.168.1.255/32       Direct  0    0          192.168.1.1      GE0/1
192.168.2.0/24         Static  60   0          0.0.0.0          Ser1/0
192.168.3.0/24         Static  60   0          0.0.0.0          Ser1/0
224.0.0.0/4            Direct  0    0          0.0.0.0          NULL0
224.0.0.0/24           Direct  0    0          0.0.0.0          NULL0
```

查看路由器 R2 的路由表：

```
[R2]disp ip routing-table

Destinations  : 19       Routes : 19

Destination/Mask       Proto   Pre  Cost       nexthop          Interface
0.0.0.0/32             Direct  0    0          127.0.0.1        InLoop0
127.0.0.0/8            Direct  0    0          127.0.0.1        InLoop0
```

```
127.0.0.0/32           Direct   0    0    127.0.0.1        InLoop0
127.0.0.1/32           Direct   0    0    127.0.0.1        InLoop0
127.255.255.255/32     Direct   0    0    127.0.0.1        InLoop0
172.16.2.0/30          Direct   0    0    172.16.2.2       Ser2/0
172.16.2.0/32          Direct   0    0    172.16.2.2       Ser2/0
172.16.2.1/32          Direct   0    0    172.16.2.1       Ser2/0
172.16.2.2/32          Direct   0    0    127.0.0.1        InLoop0
172.16.2.3/32          Direct   0    0    172.16.2.2       Ser2/0
192.168.1.0/24         Static   60   0    0.0.0.0          Ser2/0
192.168.2.0/24         Direct   0    0    192.168.2.1      GE0/0
192.168.2.0/32         Direct   0    0    192.168.2.1      GE0/0
192.168.2.1/32         Direct   0    0    127.0.0.1        InLoop0
192.168.2.255/32       Direct   0    0    192.168.2.1      GE0/0
192.168.3.0/24         Static   60   0    172.16.2.1       Ser2/0
224.0.0.0/4            Direct   0    0    0.0.0.0          NULL0
224.0.0.0/24           Direct   0    0    0.0.0.0          NULL0
255.255.255.255/32     Direct   0    0    127.0.0.1        InLoop0
```

查看路由器 R3 的路由表：

```
[R3]disp ip routing-table

Destinations  : 23      Routes : 23

Destination/Mask       Proto    Pre  Cost  nexthop         Interface
0.0.0.0/32             Direct   0    0     127.0.0.1       InLoop0
127.0.0.0/8            Direct   0    0     127.0.0.1       InLoop0
127.0.0.0/32           Direct   0    0     127.0.0.1       InLoop0
127.0.0.1/32           Direct   0    0     127.0.0.1       InLoop0
127.255.255.255/32     Direct   0    0     127.0.0.1       InLoop0
172.16.1.0/24          Direct   0    0     172.16.1.1      Ser1/0
172.16.1.0/32          Direct   0    0     172.16.1.1      Ser1/0
172.16.1.1/32          Direct   0    0     127.0.0.1       InLoop0
172.16.1.2/32          Direct   0    0     172.16.1.2      Ser1/0
172.16.1.255/32        Direct   0    0     172.16.1.1      Ser1/0
172.16.2.0/30          Direct   0    0     172.16.2.1      Ser2/0
172.16.2.0/32          Direct   0    0     172.16.2.1      Ser2/0
172.16.2.1/32          Direct   0    0     127.0.0.1       InLoop0
172.16.2.2/32          Direct   0    0     172.16.2.2      Ser2/0
172.16.2.3/32          Direct   0    0     172.16.2.1      Ser2/0
192.168.2.0/24         Static   60   0     0.0.0.0         Ser2/0
192.168.3.0/24         Direct   0    0     192.168.3.1     GE0/0
192.168.3.0/32         Direct   0    0     192.168.3.1     GE0/0
192.168.3.1/32         Direct   0    0     127.0.0.1       InLoop0
192.168.3.255/32       Direct   0    0     192.168.3.1     GE0/0
224.0.0.0/4            Direct   0    0     0.0.0.0         NULL0
224.0.0.0/24           Direct   0    0     0.0.0.0         NULL0
```

```
255.255.255.255/32    Direct  0   0         127.0.0.1            InLoop0
```

经过查看，路由器 R1 上的 3 条静态路由全被设置为送出接口，即 S1/0 接口，没有错误；路由器 R2 上有 2 条静态路由，分别指向路由器 R1 和路由器 R3 的内网网段，指向路由器 R1 的路由被设置为送出接口，没有错误，指向路由器 R3 的路由被设置为下一跳地址，也没有错误；路由器 R3 上有 1 条静态路由，指向路由器 R2 的内网网段，被设置为送出接口，没有错误。

4）查看是否为指向的目标网络地址或子网掩码错误

继续查看路由表。经过查看，目标网络地址与子网掩码也没有错误，与规划时的一致。

5）查看是否仅设置了单向路由，而没有设置双向路由

继续查看路由表。经过查看，路由器 R3 上面没有路由器 R1 的静态路由，只有路由器 R2 的静态路由。因为路由必须是双向的才能进行通信，所以 PC1 与 PC7 之间是 ping 不通的。接下来在路由器 R3 上配置到达 R1 的路由：

```
[R3]ip route-static 192.168.1.0 24 s1/0
[R3]disp ip routing-table

Destinations : 24      Routes : 24

Destination/Mask       Proto   Pre  Cost    nexthop              Interface
0.0.0.0/32             Direct  0    0       127.0.0.1            InLoop0
127.0.0.0/8            Direct  0    0       127.0.0.1            InLoop0
127.0.0.0/32           Direct  0    0       127.0.0.1            InLoop0
127.0.0.1/32           Direct  0    0       127.0.0.1            InLoop0
127.255.255.255/32     Direct  0    0       127.0.0.1            InLoop0
172.16.1.0/24          Direct  0    0       172.16.1.1           Ser1/0
172.16.1.0/32          Direct  0    0       172.16.1.1           Ser1/0
172.16.1.1/32          Direct  0    0       127.0.0.1            InLoop0
172.16.1.2/32          Direct  0    0       172.16.1.2           Ser1/0
172.16.1.255/32        Direct  0    0       172.16.1.1           Ser1/0
172.16.2.0/30          Direct  0    0       172.16.2.1           Ser2/0
172.16.2.0/32          Direct  0    0       172.16.2.1           Ser2/0
172.16.2.1/32          Direct  0    0       127.0.0.1            InLoop0
172.16.2.2/32          Direct  0    0       172.16.2.2           Ser2/0
172.16.2.3/32          Direct  0    0       172.16.2.1           Ser2/0
192.168.1.0/24         Static  60   0       0.0.0.0              Ser1/0
192.168.2.0/24         Static  60   0       0.0.0.0              Ser2/0
192.168.3.0/24         Direct  0    0       192.168.3.1          GE0/0
192.168.3.0/32         Direct  0    0       192.168.3.1          GE0/0
192.168.3.1/32         Direct  0    0       127.0.0.1            InLoop0
192.168.3.255/32       Direct  0    0       192.168.3.1          GE0/0
224.0.0.0/4            Direct  0    0       0.0.0.0              NULL0
224.0.0.0/24           Direct  0    0       0.0.0.0              NULL0
255.255.255.255/32     Direct  0    0       127.0.0.1            InLoop0
```

加上这条路由以后，PC4 与 PC1、PC7 之间便可以相互通信了，测试结果如下：

```
<H3C>ping 192.168.1.3
Ping 192.168.1.3 (192.168.1.3): 56 data bytes, press CTRL_C to break
56 bytes from 192.168.1.3: icmp_seq=0 ttl=252 time=4.000 ms
56 bytes from 192.168.1.3: icmp_seq=1 ttl=252 time=4.000 ms
56 bytes from 192.168.1.3: icmp_seq=2 ttl=252 time=4.000 ms
56 bytes from 192.168.1.3: icmp_seq=3 ttl=252 time=3.000 ms
56 bytes from 192.168.1.3: icmp_seq=4 ttl=252 time=4.000 ms

--- Ping statistics for 192.168.1.3 ---
5 packet(s) transmitted, 5 packet(s) received, 0.0% packet loss
round-trip min/avg/max/std-dev = 3.000/3.800/4.000/0.400 ms
<H3C>%Feb 20 14:37:57:555 2021 H3C PING/6/PING_STATISTICS: Ping statistics for
192.168.1.3: 5 packet(s) transmitted, 5 packet(s) received, 0.0% packet loss,
round-trip min/avg/max/std-dev = 3.000/3.800/4.000/0.400 ms.
<H3C>ping 192.168.3.3
Ping 192.168.3.3 (192.168.3.3): 56 data bytes, press CTRL_C to break
56 bytes from 192.168.3.3: icmp_seq=0 ttl=253 time=3.000 ms
56 bytes from 192.168.3.3: icmp_seq=1 ttl=253 time=3.000 ms
56 bytes from 192.168.3.3: icmp_seq=2 ttl=253 time=2.000 ms
56 bytes from 192.168.3.3: icmp_seq=3 ttl=253 time=3.000 ms
56 bytes from 192.168.3.3: icmp_seq=4 ttl=253 time=4.000 ms

--- Ping statistics for 192.168.3.3 ---
5 packet(s) transmitted, 5 packet(s) received, 0.0% packet loss
round-trip min/avg/max/std-dev = 2.000/3.000/4.000/0.632 ms
<H3C>%Feb 20 14:38:03:532 2021 H3C PING/6/PING_STATISTICS: Ping statistics for
192.168.3.3: 5 packet(s) transmitted, 5 packet(s) received, 0.0% packet loss,
round-trip min/avg/max/std-dev = 2.000/3.000/4.000/0.632 ms.
```

6）进行终端互联互通测试

按照如图 7-1 所示的网络拓扑结构做好静态路由并分配好 IP 地址后，进行终端互联互通测试，结果如表 7-3 所示。

表 7-3 终端互联互通测试结果

测试序号	交换机 S1	交换机 S2	交换机 S3	测试方法	预期测试结果	实际测试结果	是否发生故障
1	PC1、PC2、PC3	PC4、PC5、PC6		ping 命令	成功	成功	否
2	PC1、PC2、PC3		PC7、PC8、PC9	ping 命令	成功	成功	否
3		PC4、PC5、PC6	PC7、PC8、PC9	ping 命令	成功	成功	否

7）整理新的配置文档

在故障排除后，保存所有路由器的配置信息，并更新书面的记录材料，确保书面文档与实际配置保持一致，以确保下次配置正常使用。

7.3 相关知识准备

知识准备

为了能够深入地分析故障点,读者应了解静态路由的相关知识。

1)什么是静态路由

静态路由是指由网络管理员手动配置的路由信息。当网络的拓扑结构或链路状态发生变化时,网络管理员需要手动修改路由表中相关的静态路由信息。静态路由信息在默认情况下在本地有效。当然,网络管理员也可以通过对路由器进行设置,使静态路由成为共享路由。静态路由一般适用于比较简单的网络环境。在这样的环境中,网络管理员可以清楚地了解网络的拓扑结构,便于设置正确的路由信息。

2)静态路由的优点

静态路由的一个优点是网络安全性强、保密性高。因为动态路由需要路由器之间频繁地交换各自的路由表,而对路由表的分析可以揭示网络的拓扑结构和网络地址等信息,所以动态路由的安全性不如静态路由的安全性强。因此,出于安全方面的考虑,当网络规模较小时,可以采用静态路由。静态路由的另一个优点是不需要使用动态路由选择协议,从而减少了路由器的日常开销,并且可以控制路由选择。

3)静态路由的缺点

在大型和复杂的网络环境中通常不宜采用静态路由。一方面,网络管理员难以全面地了解整个网络的拓扑结构;另一方面,当网络的拓扑结构或链路状态发生变化时,需要大范围地调整路由器中的静态路由信息,这一工作的难度非常大、复杂程度非常高。

4)路由原理

当 IP 子网中的一台主机 A 发送 IP 分组给同一 IP 子网的另一台主机 B 时,主机 A 直接将 IP 分组发送到网络上,主机 B 就能接收到。但是,当主机 A 要将 IP 分组发送给不同 IP 子网上的主机 C 时,需要选择一个能到达目的子网的路由器,先将 IP 分组发送给该路由器,再由该路由器负责将 IP 分组发送到目的地。如果没有找到这样的路由器,则主机 A 会将 IP 分组发送给一个被称为"默认网关(Default Gateway)"的路由器。"默认网关"是每台主机上都有的一个配置参数,是连接在同一个网络上的某台路由器接口的 IP 地址。

路由器在转发 IP 分组时,只根据 IP 分组目的 IP 地址的网络号部分选择合适的接口,从而将 IP 分组发送出去。与主机一样,路由器也需要判断接口所连接的是否是目的子网,如果是,则直接通过该接口将 IP 分组发送到网络上,否则选择下一台路由器来发送 IP 分组。路由器也有"默认网关",用来传送不知道往哪儿发送的 IP 分组。这样,通过路由器能够将知道如何传送的 IP 分组正确发送出去,将不知道如何传送的 IP 分组发送给"默认网关"路由器。这样一级一级地传送,最终将 IP 分组发送到目的地,网络将丢弃送不到目的地的 IP 分组。

目前,TCP/IP 网络全部是通过路由器互连起来的,Internet 就是成千上万个 IP 子网通过路由器互连起来的国际性网络。这种网络被称为以路由器为基础的网络(Router Based Network),形成了以路由器为节点的"网中网"。在"网中网"中,路由器不仅负责转发 IP 分组,还负责与其他路由器进行联络,共同确定"网中网"的路由选择和维护路由表。

路由动作包括两项基本内容：寻径和转发。

寻径就是判断到达目的地的最佳路径，由路由选择算法来实现。由于涉及不同的路由选择协议和路由选择算法，因此寻径相对复杂一些。为了判断最佳路径，路由选择算法必须启动并维护包含路由信息的路由表，其中路由信息因所用的路由选择算法不同而不尽相同。路由选择算法会将收集到的不同信息填入路由表，根据路由表，路由器可以确定目的网络与下一跳地址的关系，并将该关系告诉其他路由器。路由器之间相互交换信息以进行路由更新（更新后的路由表可以正确反映网络拓扑结构的变化），并由路由器根据度量来决定最佳路径。这就是路由选择协议，常用的路由选择协议有路由信息协议（RIP）、开放式最短路径优先（OSPF）协议和边界网关协议（BGP）等。

转发就是沿最佳路径发送 IP 分组。路由器首先在路由表中进行查找，判断是否知道如何将 IP 分组发送到下一个站点（路由器或主机）。如果路由器不知道如何发送 IP 分组，则通常会将该 IP 分组丢弃，否则会根据路由表的相应表项将 IP 分组发送到下一个站点。如果目的网络直接与路由器相连，则路由器会将 IP 分组直接发送到相应的接口上。这就是路由转发协议。

路由转发协议和路由选择协议是相互配合又相互独立的概念，前者使用后者维护的路由表，同时后者需要利用前者提供的功能来发布路由协议数据分组。本项目中提到的路由协议，除非特别说明，都是指路由选择协议。

7.4 项目小结

本项目主要针对静态路由协议配置进行故障排除，主要思路是先硬件、后软件。因为本书的项目都是在模拟环境下实现的，所以不考虑物理线路、电源、模块损坏等硬件问题，直接从软件开始进行故障排除，但是在实际应用环境中，需要按照故障排除规范一一检查。对于本项目，第一，检查 IP 地址规划是否错误；第二，检查静态路由协议的下一跳地址、接口配置是否有问题；第三，检查路由器或交换机的配置是否有错误（如是否有配置双向路由等），直到检查出故障为止。

素质拓展：弹指一挥 突飞猛进

中国科技的十年对比：十年弹指一挥间，中国科学技术突飞猛进地发展。天眼探空、蛟龙入海、墨子传信、超算发威、北斗组网、嫦娥奔月……一项项科技突破，见证中国科学技术从"追赶"到"并跑"，甚至"领跑"的转变。

中国科学技术的十年对比，记录着永不停步的接力奋斗，也鼓励着广大科研工作者们，永不停步、继续向前！

2011 年 1 月 11 日中午 12 时 50 分左右，歼-20 首架技术验证机在四川成都完成首次升空飞行测试。在 2021 年第十三届中国航展开幕首日现场，歼-20 双机编队完成多个高难度战术动作。歼-20以新装备、新编队、新姿态参加本届航展,这也是歼-20换装国产发动机后首次对外公开展示。在飞行表演中,多个动作充分展现了歼-20优秀的低空高速、小半径转弯及大迎角飞行能力。歼-20是一个高技术集成体,代表了中国航空科技水平的重大突破,也标志着中国航空工业已经接近世界先进水平。

增值服务

提高系统的可管理性,完善网络系统的远程维护和登录授权。从优化系统性能角度完善网络配置参数。

在解决项目故障和整理故障修复文档时,与公司信息技术人员进行沟通,了解公司部门之间业务可靠性的需求。建议在业务可靠性要求高的链路上配置双链路,以及默认路由和浮动路由,以提高链路的可靠性,确保所有计算机能够互相访问。在本项目中,可以在路由器 R1 与 R3 之间增加一条链路,并配置浮动路由,以提高网络的可靠性。

参考命令:"ip route-static 0.0.0.0 0.0.0.0 172.16.1.1 preference 100"表示通过 preference 100 参数实现配置浮动路由,以提高链路的可靠性。

7.5 课后实训

项目内容:某公司新成立了 A 和 B 两个分公司,需要在总公司与分公司 A 和分公司 B 之间建立局域网,因为网络规模不大,所以使用静态路由。总公司的内网使用 192.168.10.0/24 网段,分公司 A 和分公司 B 的内网使用 192.168.20.0/24 和 192.168.30.0/24 网段,各公司的链路之间采用 172.16.1.0/30 和 172.16.2.0/30 网段。现已完成配置,但是在测试过程中出现了故障。请根据故障现象,完成以下任务。

① 根据要求检查故障现象。
② 根据故障现象收集故障信息。
③ 利用结构化故障排除方法完成故障定位。
④ 修改故障配置并说明故障原因。
⑤ 更新配置文档。

项目 8

动态路由协议 RIP 的故障排除

内容介绍

某公司刚刚成立，共有 3 栋建筑楼，一个部门占一栋建筑楼。为了方便员工之间的信息共享与交流，需要建设一个内部局域网，要求每栋建筑楼内要有一个自己的局域网。考虑到公司人数不多，建筑楼也不多，因此可以采用 RIP 来实现各部门业务的互联互通。

任务安排

任务 1　进行 RIP 互连的网络配置
任务 2　进行网络更新过程中的故障分析与排除

学习目标

◇ 了解 RIP 常见故障的原因
◇ 掌握故障排除的思路
◇ 学会结构化故障排除方法
◇ 学会 RIP 相关故障排除及文档更新的方法

素质目标

要有健康的体魄、心理和健全的人格，乐于运动。养成良好的劳动习惯、行为习惯和工作习惯。

8.1 RIP 配置分析与实施

发现故障

网络管理员王工在接到公司的网络建设任务后,根据公司业务需求制定了网络实施方案,主要操作如下。

① 规划网络拓扑结构。
② 规划 IP 地址。
③ 规划路由器、交换机所连接的接口。
④ 对路由器、交换机、PC 终端进行配置。
⑤ 进行终端互联互通测试。

1)规划网络拓扑结构

网络管理员在分析业务需求后,规划了网络拓扑结构,如图 8-1 所示。

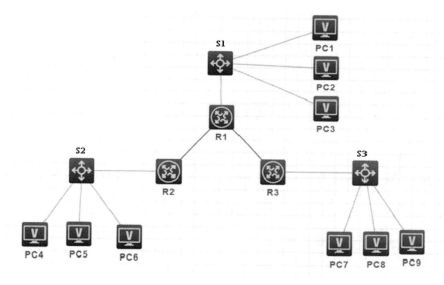

图 8-1 网络拓扑结构

在图 8-1 中,PC1、PC2、PC3 在同一个网段;PC4、PC5、PC6 在同一个网段;PC7、PC8、PC9 在同一个网段。路由器 R1 与路由器 R2 之间的互联接口是 S1/0;路由器 R1 与路由器 R3 之间的互联接口是 S2/0;路由器 R1 的 G0/1 接口与交换机 S1 的 G0/1 接口互连;路由器 R2 的 G0/1 接口与交换机 S2 的 G0/1 接口互连;路由器 R3 的 G0/1 接口与交换机 S3 的 G0/1 接口互连;PC1、PC2、PC3 分别与交换机 S1 的 G0/2、G0/3、G0/4 接口互连;PC4、PC5、PC6 分别与交换机 S2 的 G0/2、G0/3、G0/4 接口互连;PC7、PC8、PC9 分别与交换机 S3 的 G0/2、G0/3、G0/4 接口互连。

2)规划 IP 地址

参照图 8-1,使用子网划分方法进行 IP 地址规划,如表 8-1 所示。

表 8-1 各接口 IP 地址规划

路由器 R1 与路由器 R3 的互联地址	路由器 R2 与路由器 R3 的互联地址	路由器 R1 上的网关地址、业务地址	路由器 R2 上的网关地址、业务地址	路由器 R3 上的网关地址、业务地址
192.168.5.1/24	192.168.3.1/24	网关地址：192.168.1.1/24	网关地址：192.168.2.1/24	网关地址：192.168.4.1/24
192.168.5.2/24	192.168.3.2/24	业务地址：192.168.1.0/24	业务地址：192.168.2.0/24	业务地址：192.168.4.0/24

3）规划路由器、交换机所连接的接口，并对路由器、交换机、PC 终端进行相关配置。配置结果如下。查看路由器 R1 的配置：

```
[R1]disp cu
#
 version 7.1.075, Alpha 7571
#
 sysname R1
#
 rip 1
 version 2
 network 192.168.1.0
 network 192.168.3.0
 network 192.168.5.0
#
 system-working-mode standard
 xbar load-single
 password-recovery enable
 lpu-type f-series
#
vlan 1
#
interface Serial1/0
 ip address 192.168.3.1 255.255.255.0
#
interface Serial2/0
 ip address 192.168.5.1 255.255.255.0
#
interface Serial3/0
#
interface Serial4/0
#
interface NULL0
#
interface GigabitEthernet0/0
 port link-mode route
 combo enable copper
```

```
#
interface GigabitEthernet0/1
 port link-mode route
 combo enable copper
 ip address 192.168.1.1 255.255.255.0
#
interface GigabitEthernet0/2
 port link-mode route
 combo enable copper
#
interface GigabitEthernet5/0
 port link-mode route
 combo enable copper
#
interface GigabitEthernet5/1
 port link-mode route
 combo enable copper
#
interface GigabitEthernet6/0
 port link-mode route
 combo enable copper
#
interface GigabitEthernet6/1
 port link-mode route
 combo enable copper
#
 scheduler logfile size 16
#
line class aux
 user-role network-operator
#
line class console
 user-role network-admin
#
line class tty
 user-role network-operator
#
line class vty
 user-role network-operator
#
line aux 0
 user-role network-operator
#
line con 0
 user-role network-admin
```

```
#
line vty 0 63
 user-role network-operator
#
 undo info-center enable
#
domain name system
#
 domain default enable system
#
role name level-0
 description Predefined level-0 role
#
role name level-1
 description Predefined level-1 role
#
role name level-2
 description Predefined level-2 role
#
role name level-3
 description Predefined level-3 role
#
role name level-4
 description Predefined level-4 role
#
role name level-5
 description Predefined level-5 role
#
role name level-6
 description Predefined level-6 role
#
role name level-7
 description Predefined level-7 role
#
role name level-8
 description Predefined level-8 role
#
role name level-9
 description Predefined level-9 role
#
role name level-10
 description Predefined level-10 role
#
role name level-11
 description Predefined level-11 role
```

```
#
role name level-12
 description Predefined level-12 role
#
role name level-13
 description Predefined level-13 role
#
role name level-14
 description Predefined level-14 role
#
user-group system
#
return
```

查看路由器 R2 的配置：

```
[R2]disp cu
#
 version 7.1.075, Alpha 7571
#
 sysname R2
#
rip 1
 network 192.168.2.0
 network 192.168.3.0
 silent-interface Serial1/0
#
 system-working-mode standard
 xbar load-single
 password-recovery enable
 lpu-type f-series
#
vlan 1
#
interface Serial1/0
 ip address 192.168.3.2 255.255.255.0

#
interface Serial2/0
 ip address 172.16.2.1 255.255.255.0
#
interface Serial3/0
#
interface Serial4/0
#
interface NULL0
#
```

```
interface GigabitEthernet0/0
 port link-mode route
 combo enable copper
#
interface GigabitEthernet0/1
 port link-mode route
 combo enable copper
 ip address 192.168.2.1 255.255.255.0
#
interface GigabitEthernet0/2
 port link-mode route
 combo enable copper
#
interface GigabitEthernet5/0
 port link-mode route
 combo enable copper
#
interface GigabitEthernet5/1
 port link-mode route
 combo enable copper
#
interface GigabitEthernet6/0
 port link-mode route
 combo enable copper
#
interface GigabitEthernet6/1
 port link-mode route
 combo enable copper
#
 scheduler logfile size 16
#
line class aux
 user-role network-operator
#
line class console
 user-role network-admin
#
line class tty
 user-role network-operator
#
line class vty
 user-role network-operator
#
line aux 0
 user-role network-operator
```

```
#
line con 0
 user-role network-admin
#
line vty 0 63
 user-role network-operator
#
 undo info-center enable
#
domain name system
#
 domain default enable system
#
role name level-0
 description Predefined level-0 role
#
role name level-1
 description Predefined level-1 role
#
role name level-2
 description Predefined level-2 role
#
role name level-3
 description Predefined level-3 role
#
role name level-4
 description Predefined level-4 role
#
role name level-5
 description Predefined level-5 role
#
role name level-6
 description Predefined level-6 role
#
role name level-7
 description Predefined level-7 role
#
role name level-8
 description Predefined level-8 role
#
role name level-9
 description Predefined level-9 role
#
role name level-10
```

```
 description Predefined level-10 role
#
role name level-11
 description Predefined level-11 role
#
role name level-12
 description Predefined level-12 role
#
role name level-13
 description Predefined level-13 role
#
role name level-14
 description Predefined level-14 role
#
user-group system
#
return
```

查看路由器 R3 的配置：

```
[R3]disp cu
#
 version 7.1.075, Alpha 7571
#
 sysname R3
#
rip 1
 network 192.168.5.0
 network 192.168.40.0
#
 system-working-mode standard
 xbar load-single
 password-recovery enable
 lpu-type f-series
#
vlan 1
#
interface Serial1/0
 ip address 172.16.4.1 255.255.255.0
#
interface Serial2/0
 ip address 192.168.5.2 255.255.255.0
#
interface Serial3/0
#
interface Serial4/0
```

```
#
interface NULL0
#
interface GigabitEthernet0/0
 port link-mode route
 combo enable copper
#
interface GigabitEthernet0/1
 port link-mode route
 combo enable copper
 ip address 192.168.4.1 255.255.255.0
#
interface GigabitEthernet0/2
 port link-mode route
 combo enable copper
#
interface GigabitEthernet5/0
 port link-mode route
 combo enable copper
#
interface GigabitEthernet5/1
 port link-mode route
 combo enable copper
#
interface GigabitEthernet6/0
 port link-mode route
 combo enable copper
#
interface GigabitEthernet6/1
 port link-mode route
 combo enable copper
#
scheduler logfile size 16
#
line class aux
 user-role network-operator
#
line class console
 user-role network-admin
#
line class tty
 user-role network-operator
#
line class vty
```

```
 user-role network-operator
#
line aux 0
 user-role network-operator
#
line con 0
 user-role network-admin
#
line vty 0 63
 user-role network-operator
#
 undo info-center enable
#
domain name system
#
 domain default enable system
#
role name level-0
 description Predefined level-0 role
#
role name level-1
 description Predefined level-1 role
#
role name level-2
 description Predefined level-2 role
#
role name level-3
 description Predefined level-3 role
#
role name level-4
 description Predefined level-4 role
#
role name level-5
 description Predefined level-5 role
#
role name level-6
 description Predefined level-6 role
#
role name level-7
 description Predefined level-7 role
#
role name level-8
 description Predefined level-8 role
#
```

```
role name level-9
 description Predefined level-9 role
#
role name level-10
 description Predefined level-10 role
#
role name level-11
 description Predefined level-11 role
#
role name level-12
 description Predefined level-12 role
#
role name level-13
 description Predefined level-13 role
#
role name level-14
 description Predefined level-14 role
#
user-group system
#
return
```

交换机和 PC 终端的相关配置可参考本书配套素材，这里不进行细述。

4）进行终端互联互通测试

按照图 8-1 分配好地址并做好动态路由以后，进行终端互联互通测试，结果如表 8-2 所示。

表 8-2　终端互联互通测试结果

测试序号	交换机 S1	交换机 S2	交换机 S3	测试方法	预期测试结果	实际测试结果	是否发生故障
1	PC1、PC2、PC3	PC4、PC5、PC6		ping 命令	成功	失败	是
2	PC1、PC2、PC3		PC7、PC8、PC9	ping 命令	成功	失败	是
3		PC4、PC5、PC6	PC7、PC8、PC9	ping 命令	成功	失败	是

由表 8-2 可知，有 3 项测试失败，因此本次操作没有成功完成项目目标。这是什么原因造成的故障现象呢？是规划设计的问题、操作的问题，还是 RIP 概念没有理解清楚的问题？据此，要深入进行故障分析，以确定问题所在。

8.2　RIP 配置故障分析与排除

排除故障

1. 故障分析方法

根据结构化故障排除思路，故障排除需要遵循 OSI 参考模型的要求。这里采用自上而下的方法进行排除，同时使用与动态路由协议相关的技术进行综合分析。

从故障排除理论体系角度进行分析，故障点可能包括以下几方面：IP 地址配置错误、RIP 版本错误、RIP 配置错误。

2. 分析故障点

从 RIP 的实现及技术原理进行分析，可能存在以下故障点。

① RIP 版本不一致，如一台路由器用的是 RIPv1，另一台路由器用的是 RIPv2。
② 发布的网络地址错误。
③ RIP 配置错误。

由于本次故障主要通过模拟环境来实现，因此可以忽略物理问题及设备问题。本次主要从 RIP 的相关概念及 RIP 的配置和操作方面进行故障排除。

本次故障排除任务的目标如下。
① 查看 IP 地址配置，确定已有的配置信息。
② 查看 RIP 配置，确定已有的配置信息。

完成本次故障排除任务所需的设备如下。
① 一台装有超级终端软件或 Telnet 软件的计算机，同时确定访问所需的用户名和口令。
② 配置线缆。
③ 笔和纸，用于记录相关信息。

1）查看 IP 地址配置

查看路由器 R1 的 IP 地址配置：

```
[R1]disp ip in brief
*down: administratively down
(s): spoofing (l): loopback
Interface            Physical     Protocol     IP Address       Description
GE0/0                down         down         --               --
GE0/1                up           up           192.168.1.1      --
GE0/2                down         down         --               --
GE5/0                down         down         --               --
GE5/1                down         down         --               --
GE6/0                down         down         --               --
GE6/1                down         down         --               --
Ser1/0               up           up           192.168.3.1      --
Ser2/0               up           up           192.168.5.1      --
Ser3/0               down         down         --               --
Ser4/0               down         down         --               --
```

查看路由器 R2 的 IP 地址配置：

```
[R2]disp ip in brief
*down: administratively down
(s): spoofing (l): loopback
Interface            Physical     Protocol     IP Address       Description
```

Interface	Physical	Protocol	IP Address	Description
GE0/0	down	down	--	--
GE0/1	up	up	192.168.2.1	--
GE0/2	down	down	--	--
GE5/0	down	down	--	--
GE5/1	down	down	--	--
GE6/0	down	down	--	--
GE6/1	down	down	--	--
Ser1/0	up	up	192.168.3.2	--
Ser2/0	down	down	172.16.2.1	--
Ser3/0	down	down	--	--
Ser4/0	down	down	--	--

查看路由器 R3 的 IP 地址配置：

```
[R3]disp ip in brief
*down: administratively down
(s): spoofing  (l): loopback
```

Interface	Physical	Protocol	IP Address	Description
GE0/0	down	down	--	--
GE0/1	up	up	192.168.4.1	--
GE0/2	down	down	--	--
GE5/0	down	down	--	--
GE5/1	down	down	--	--
GE6/0	down	down	--	--
GE6/1	down	down	--	--
Ser1/0	down	down	172.16.4.1	--
Ser2/0	up	up	192.168.5.2	--
Ser3/0	down	down	--	--
Ser4/0	down	down	--	--

经过查看，路由器的 IP 地址配置都没有出现错误，因此可以排除 IP 地址错误问题。

2）查看接口配置

经过查看，路由器 R1、路由器 R2、路由器 R3 的接口配置都是 up 状态，协议也是 up 状态，并且与规划时所用的接口一致，因此可以排除接口故障。

3）查看 RIP 版本是否一致

查看路由器 R1 的 RIP 版本：

```
[R1]disp rip
 Public VPN-instance name:
   RIP process: 1
     RIP version: 2
     Preference: 100
     Checkzero: Enabled
     Default cost: 0
     Summary: Enabled
     Host routes: Enabled
     Maximum number of load balanced routes: 32
     Update time    :   30 secs  Timeout time           :  180 secs
```

```
        Suppress time :  120 secs  Garbage-collect time :  120 secs
        Update output delay:   20(ms) Output count:    3
        TRIP retransmit time:    5(s)  Retransmit count: 36
        Graceful-restart interval:   60 secs
        Triggered Interval : 5 50 200
        BFD: Disabled
        Silent interfaces: None
        Default routes: Disabled
        Verify-source: Enabled
        Networks:
            192.168.1.0           192.168.3.0
            192.168.5.0
        Configured peers: None
        Triggered updates sent: 3
        Number of routes changes: 3
        Number of replies to queries: 0
```

查看路由器 R2 的 RIP 版本：

```
[R2]disp rip
 Public VPN-instance name:
   RIP process: 1
      RIP version: 1
      Preference: 100
      Checkzero: Enabled
      Default cost: 0
      Summary: Enabled
      Host routes: Enabled
      Maximum number of load balanced routes: 32
      Update time   :   30 secs  Timeout time      :  180 secs
      Suppress time :  120 secs  Garbage-collect time :  120 secs
      Update output delay:   20(ms) Output count:    3
      TRIP retransmit time:    5(s)  Retransmit count: 36
      Graceful-restart interval:   60 secs
      Triggered Interval : 5 50 200
      BFD: Disabled
      Silent interfaces:
          Ser1/0
      Default routes: Disabled
      Verify-source: Enabled
      Networks:
          192.168.2.0           192.168.3.0
      Configured peers: None
      Triggered updates sent: 3
      Number of routes changes: 4
      Number of replies to queries: 0
```

查看路由器 R3 的 RIP 版本：

```
[R3]disp rip
 Public VPN-instance name:
  RIP process: 1
    RIP version: 1
    Preference: 100
    Checkzero: Enabled
    Default cost: 0
    Summary: Enabled
    Host routes: Enabled
    Maximum number of load balanced routes: 32
    Update time    :   30 secs  Timeout time       :  180 secs
    Suppress time  :  120 secs  Garbage-collect time :  120 secs
    Update output delay:    20(ms)  Output count:     3
    TRIP retransmit time:    5(s)  Retransmit count: 36
    Graceful-restart interval:    60 secs
    Triggered Interval : 5 50 200
    BFD: Disabled
    Silent interfaces: None
    Default routes: Disabled
    Verify-source: Enabled
    Networks:
        192.168.5.0              192.168.40.0
    Configured peers: None
    Triggered updates sent: 0
    Number of routes changes: 3
    Number of replies to queries: 0
```

经过查看，路由器 R1 的 RIP 版本为第 2 版，路由器 R2、路由器 R3 的 RIP 版本为第 1 版，而第 1 版和第 2 版 RIP 不兼容，所以会出现故障。将路由器 R1、路由器 R2、路由器 R3 的 RIP 版本修改一致，再次进行终端互联互通测试，结果如图 8-2 所示。

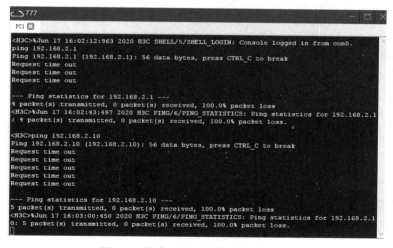

图 8-2　终端互联互通测试结果（1）

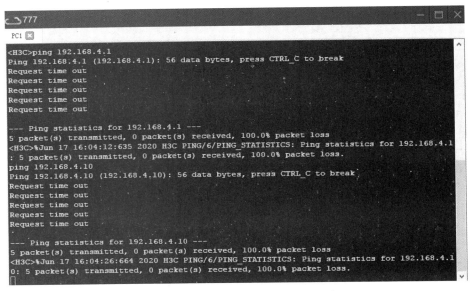

图 8-2 终端互联互通测试结果（1）（续）

经过测试，网络还是不通，接下来进行进一步分析。

4）查看发布的网络地址是否正确

与步骤 3）中使用的命令一样，这里也使用 disp rip 命令查看网络地址，结果如下。

查看路由器 R1 的网络地址：

```
[R1]
Networks:
        192.168.1.0          192.168.3.0
        192.168.5.0
```

查看路由器 R2 的网络地址：

```
[R2]
Networks:
        192.168.2.0          192.168.3.0
```

查看路由器 R3 的网络地址：

```
[R3]
Networks:
        192.168.5.0          192.168.40.0
```

经过查看，并结合表 8-1 中的 IP 地址规划，可知路由器 R1 和路由器 R2 的接口 IP 地址、协议版本号及协议发布的网段都是正确的，路由器 R3 的 RIP 中有一个错误的网段 192.168.40.0/24，而正确的网段应该是 192.168.4.0/24。修改这个错误，再次进行终端互联互通测试，结果如图 8-3 所示。

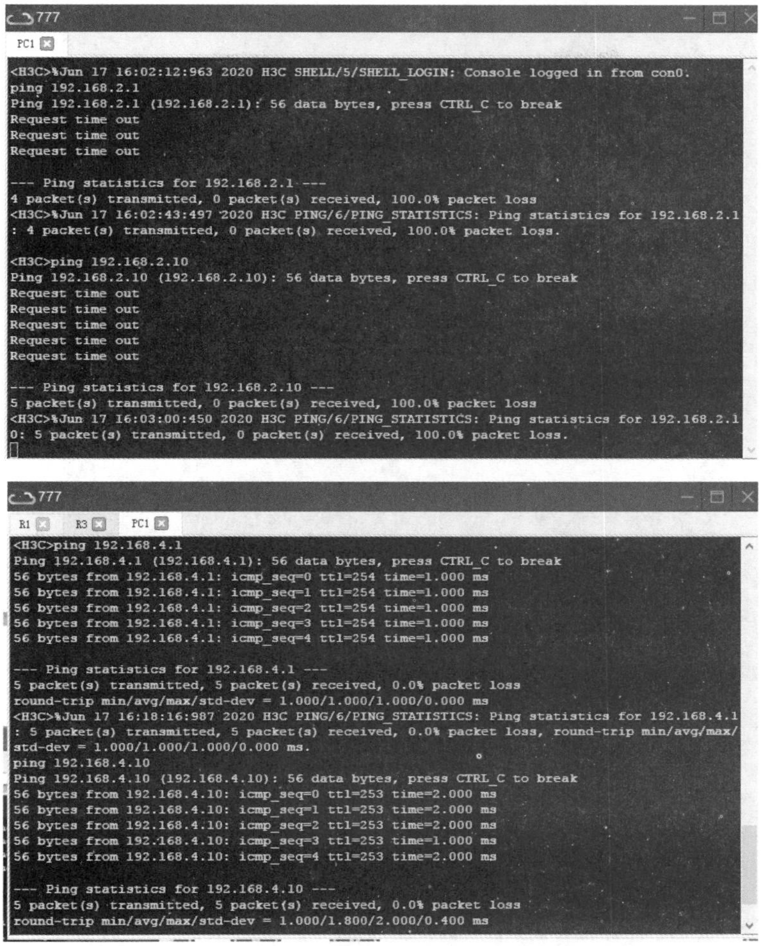

图 8-3　终端互联互通测试结果（2）

经过测试，路由器 R1 的网络 ping 不通路由器 R2 所在的网络，但是能 ping 通路由器 R3 所在的网络；路由器 R2 的网络 ping 不通路由器 R3 所在的网络。为什么会这样呢？接下来继续进行分析。

5）查看 RIP 配置是否正确

查看路由器 R1 的 RIP 配置：

```
rip 1
 version 2
 network 192.168.1.0
 network 192.168.3.0
    network 192.168.5.0
```

查看路由器 R2 的 RIP 配置：

```
rip 1
 network 192.168.2.0
 network 192.168.3.0
```

```
silent-interface Serial1/0
```
查看路由器 R3 的 RIP 配置：
```
rip 1
 network 192.168.5.0
   network 192.168.40.0
```
经过查看，在路由器 R2 的 RIP 配置中，S1/0 接口被设置为了 silent，所以路由器 R2 收到更新消息也不理会，导致网络不通。修改这个错误，再次进行网络终端互联互通测试，结果如图 8-4 所示。

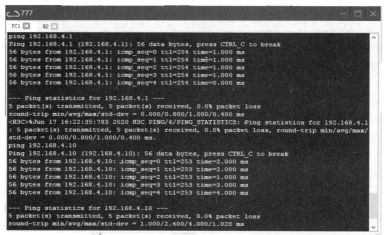

图 8-4 终端互联互通测试结果（3）

经过测试，现在所有的网络都可以互联互通了，故障排除完成。

6）整理新的配置文档

在故障排除后，保存所有路由器的配置信息，并更新书面的记录材料，确保书面文档和实际配置保持一致，以确保下次配置正常使用。

8.3 相关知识准备

知识准备

为了能够深入地分析故障点，读者应了解 RIP 的相关知识。

1）什么是 RIP

RIP（Routing Information Protocol，路由信息协议）是一种距离矢量路由协议，最大的特点是无论实现原理还是配置方法都比较简单。

2）RIP 的优点

RIP 的优点是配置简单，适用于规模较小的网络。

3）RIP 的缺点

① 会产生大量的广播。RIP 向所有邻居每隔 30s 广播一次完整的路由表，这样将大量占用宝贵的带宽资源。

② 没有成本概念。RIP 没有网络延迟和链路成本的概念，当采用 RIP 时，路由/转发的决定仅基于跳数，这样很容易导致无法选择最佳路由。

③ 支持的网络规模有限。由于 RIP 最多支持 15 跳，当跳数为 16 时，网络将认为无法到达目的地，因此 RIP 只适用于规模较小的网络。

④ 安全性差。RIP 接受来自任何设备的路由更新。

4）RIP 的特点

① RIP 是一种距离矢量路由协议。

② RIP 使用跳数作为路由选择的度量。

③ 当到达目的网络的跳数超过 15 跳时，数据包将被丢掉。

④ 在默认情况下，RIP 路由更新广播周期为 30s。

5）RIP 的工作原理

RIP 是基于距离矢量的算法。

在实现路由时，RIP 作为一个系统常驻进程而存在于路由器中，负责接收网络系统中其他路由器的路由信息，以动态维护本地网络层路由表。RIP 使用跳数来衡量到达目标地址的路由距离。每台路由器在向相邻路由器发出路由信息时，每经过一台路由器，跳数就会加 1。

（1）路由更新。

RIP 中路由的更新是通过定时广播实现的。在默认情况下，路由器每隔 30s 就向与它相连的网络广播自己的路由表，接到广播的路由器会将收到的信息添加至自己的路由表中。每台路由器都如此广播，最终网络上所有的路由器都会获得全部的路由信息。在正常情况下，路由器每隔 30s 就可以收到一次路由信息确认。如果经过 180s，即 6 个更新周期，任何路由项都没有得到确认，那么路由器会认为它已经失效了；如果经过 240s，即 8 个更新周期，路由项仍没有得到确认，那么它会被从路由表中删除。上面的 30s、180s 和 240s 的延时都是由计时器控制的，它们分别是更新计时器（Update Timer）、无效计时器（Invalid Timer）和刷新计时器（Flush Timer）。

（2）路由循环。

距离矢量类的算法容易产生路由循环，RIP 是距离矢量算法中的一种，所以它也不例外。如果网络中存在路由循环，那么信息会被循环传递，永远不能到达目的地。为了避免这个问题，RIP 等距离矢量算法实现了下面 4 种机制。

① 水平分割（Split Horizon）。水平分割可以保证路由器能够记住每一条路由信息的来源，并且不在收到这条信息的接口上再次发送它。这是保证不产生路由循环的最基本措施。

② 毒性逆转（Poison Reverse）。当一条路径信息变为无效之后，路由器不会立即将这条信息从路由表中删除，而是使用度量值 16（表示不可达）将这条信息广播出去。这样虽然会增加路由表的大小，但对消除路由循环很有帮助。毒性逆转可以立即清除相邻路由器之间的任何环路。

③ 触发更新（Trigger Update）。当路由表发生变化时，立即将更新报文广播给相邻的所有路由器，而不是等待 30s 的更新周期。同样，当一台路由器刚启动 RIP 时，会发送更新广播来请求报文，收到此广播的相邻路由器会立即应答一个更新报文，而不必等到下一个更新周期。这样，网络拓扑结构的变化会迅速在网络中传播，从而降低产生路由循环的可能性。

④ 抑制计时（Holddown Timer）。当一条路由信息无效之后，在一段时间内这条路由会进入抑制状态，即在一定时间内不再接收同一目的地址的路由更新。如果路由器从一个网段上得知某条路由失效，却立即在另一个网段上得知这条路由有效，那么这个有效的信息往往是不正确的。抑制计时可以防止发生这种情况。此外，当一条链路频繁起停时，抑制计时可以减少路由的浮动，提高网络的稳定性。

即便采用了上面的 4 种机制，也不能完全解决路由循环的问题。一旦出现路由循环，路由项的度量值就会不断增加，直至无穷大（Count to Infinity）。这是因为路由信息在循环传递过程中，每经过一台路由器，度量值就会增加 1。当度量值增加到 16 时，路径会被视为不可达。

8.4 项目小结

本项目主要针对 RIP 配置进行故障排除，主要思路是先硬件、后软件。因为本书的项目都是在模拟环境下实现的，所以直接从软件开始进行故障排除。对于本项目，第一，检查 IP 地址规划是否错误；第二，检查 RIP 是否开启；第三，检查是否已启用相应的网段；第四，检查发布的网段的子网掩码是否正确；第五，检查版本差异问题，如 RIPv1 与 RIPv2 不兼容问题；第六，检查各接口情况，直到检查出故障为止。

素质拓展：时不我待 自我创新

中国 TSN 芯片进入商用领域。

工业和信息化部工业互联网产业联盟于 2022 年 9 月公布"时间敏感网络（TSN）产业链名录计划"，中国自主设计的 TSN 芯片被列入该名录。

TSN 是实现全球工业控制、汽车控制、飞机控制等工业网络通信协议及标准统一的国际标准技术。随着 5G 物联网、工业互联网等新一代信息通信技术的发展，TSN 具有的确定性和微秒级交互特性，这些特性增强了无人驾驶、边缘计算、虚拟现实等技术的现实应用，推动了传统离散工业的数字化转型和智能制造的升级。

增值服务

党的二十大报告提出了继续推进理论创新的科学方法，准确把握"六个必须坚持"。

随着网络规模的扩大及网络稳定性要求的提高，我们要聚焦实践中遇到的新问题，分析网络协议发展进程中的突出问题，把握技术发展趋势。在网络规模不断扩展的情况下，建议优先选用 OSPF，或者对原有网络的 RIP 配置进行优化，同时使用 BFD 联动特性，以提高 RIP 的收敛速度，实现链路的快速切换。采用新理念、新思路、新办法解决问题，不断创新网络应用，提高实效。

8.5 课后实训

项目内容：某公司最近并购了两家小公司，现在要将这两家公司的局域网通过路由器连接到公司的出口路由器上，并在路由器之间配置动态路由协议 RIP，以实现公司内部主机之间的互通。总公司内部网络选用 10.10.10.0/24 网段，网关选用网段最后一个 IP 地址；两家分公司内部网络分别选用 10.10.20.0/24 网段和 10.10.30.0/24 网段，网关选用网段最后一个 IP 地址；公司路由器之间的链路地址选用 172.16.1.0/30 网段。

在完成网络配置后进行测试时，网络出现了故障现象。请按如下操作步骤排除故障。

① 根据要求检查故障现象。
② 根据故障现象收集故障信息。
③ 利用结构化故障排除方法完成故障定位。
④ 修改故障配置并说明故障原因。
⑤ 更新配置文档。

项目 9

动态路由协议 OSPF 的故障排除

内容介绍

某公司的两家子公司要和公司总部网络实现互联,要求在确保子公司可以访问公司总部的网站服务器的同时,可以在一定程度上减轻路由器的负担。在经过综合分析后,该公司决定采用 OSPF 来实现各部门业务的互联互通。

任务安排

任务 1　进行 OSPF 互联的网络配置
任务 2　进行网络更新过程中的故障分析与排除

学习目标

◇ 了解 OSPF 常见故障的产生原因
◇ 掌握故障排除的思路
◇ 学会结构化故障排除方法
◇ 学会 OSPF 相关故障排除及文档更新的方法

素质目标

勤于思考,知行合一,在实践中提高专业技能和职业素养。

项目 9　动态路由协议 OSPF 的故障排除

9.1　OSPF 配置分析与实施

发现故障

网络管理员王工在接到公司的网络建设任务后，根据公司业务需求，制定了网络实施方案，主要操作如下。

① 为总部和两家子公司的路由器规划 IP 地址，并规划各路由器的互联接口。
② 按规划进行配置并测试网络连通性。
③ 查看现有路由器的配置。
④ 根据相关知识点判断故障点的位置，并排除故障。
⑤ 进行终端互联互通测试。

公司网络拓扑结构如图 9-1 所示。

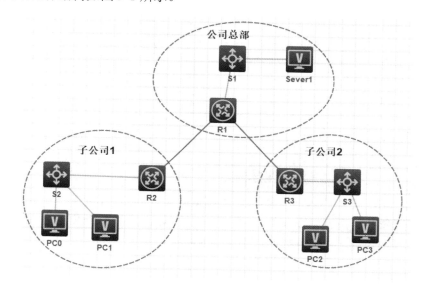

图 9-1　公司网络拓扑结构

1）规划各路由器的互联接口及 IP 地址

各接口及主机的 IP 地址规划如表 9-1 所示。

表 9-1　各接口及主机的 IP 地址规划

路由器	接口及主机	IP 地址
R1	G0/1	222.4.10.1
	S1/0	222.4.0.1
	S2/0	222.4.0.6
	Server1	222.4.10.2
R2	S1/0	222.4.0.2
	G0/1	192.168.1.1
	PC0	192.168.1.2
	PC1	192.168.1.3

续表

路由器	接口及主机	IP 地址
R3	S2/0	222.4.0.5
	G0/1	192.168.2.1
	PC2	192.168.2.2
	PC3	192.168.2.3

2）按规划进行配置并进行终端互联互通测试

在按照表 9-1 分配地址并进行动态路由后，进行终端互联互通测试，结果如表 9-2 所示。

表 9-2　终端互联互通测试结果

测试序号	路由器 R1	路由器 R2	路由器 R3	测试方法	预期测试结果	实际测试结果	是否发生故障
1	Server1	PC0		web 访问	成功	失败	是
2		PC1 和 R2-G0/1		ping 命令	成功	成功	否
3	Server1		PC2	web 访问	成功	失败	是
4			PC3 和 R3-G0/1	ping 命令	成功	成功	否
5	R1-S1/0	PC1		ping 命令	成功	失败	是
6	R1-S2/0		PC3	ping 命令	成功	失败	是
7	Server1 和 R1-G0/1			ping 命令	成功	成功	否
8	R1-S1/0	R2-S2/0		ping 命令	成功	成功	否
9	R1-S2/0		R3-S2/0	ping 命令	成功	成功	否

由表 9-2 可知，有 4 项测试失败，因此本次操作没有成功完成项目目标。这是什么原因造成的故障现象呢？是规划设计的问题、操作的问题，还是 OSPF 概念没有理解清楚的问题？据此，要深入进行故障分析，以确定问题所在。

9.2　OSPF 配置故障分析与排除

排除故障

1．故障分析方法

根据结构化故障排除思路，故障排除需要遵循 OSI 参考模型的要求。这里采用自下而上的方法，从物理层往上依次进行排除，同时利用命令和 OSPF 原理进行综合分析。

从故障排除理论体系角度进行分析，故障点可能包括以下几方面：物理问题、损坏的电缆、损坏的接口、电源故障、协议错误、性能问题、配置错误。

2．分析故障点

从 OSPF 的实现及技术原理进行分析，可能存在以下故障点。

① 错误的 OSPF 区域号。

② OSPF 发布的路由配置错误。

③ 错误的被动接口（Passive-Interface，PI）设置。

④ 网络中有的路由器 ID 相同。

由于本次故障主要在模拟环境下实现，因此可以忽略物理问题及设备问题。本次主要从 OSPF 的相关概念及 OSPF 的配置和操作方面进行故障排除。

本次故障排除任务的目标如下。

① 查看 IP 地址配置，确定已有的配置信息。

② 查看 OSPF 配置，确定已有的配置信息。

完成本次故障排除任务所需的设备如下。

① 一台装有超级终端软件或 Telnet 软件的计算机，同时确定访问所需的用户名和口令。

② 配置线缆。

③ 笔和纸，用于记录相关信息。

1）查看现有路由器的配置

① 使用超级终端软件或 Telnet 软件连接交换机，并使用 display ospf peer brief 命令查看路由器 R1、R2 和 R3 的邻居信息。

查看路由器 R1 的邻居信息：
```
R1# display ospf peer brief
R1#
```

查看路由器 R2 的邻居信息：
```
R2# display ospf peer brief
R2#
```

查看路由器 R3 的邻居信息：
```
R3# display ospf peer brief
R3#
```

② 整理路由器 R1、R2 和 R3 的邻居信息，如表 9-3 所示。

表 9-3 路由器的邻居信息

路由器	OSPF 邻居路由器
R1	无
R2	无
R3	无

③ 使用 disp cu 命令查看路由器 R1、R2 和 R3 的 OSPF 发布的路由及被动接口。

查看路由器 R1 的 OSPF 发布的路由及被动接口：
```
[R1]disp cu
#
 version 7.1.075, Alpha 7571
#
 sysname R1
#
ospf 1
 area 0.0.0.0
  network 222.4.0.0 0.0.0.3
```

```
 network 222.4.0.4 0.0.0.3
 network 222.4.10.0 0.0.0.255
#
 system-working-mode standard
 xbar load-single
 password-recovery enable
 lpu-type f-series
#
vlan 1
#
interface Serial1/0
 IP address 222.4.0.1 255.255.255.252
#
interface Serial2/0
 IP address 222.4.0.6 255.255.255.252
#
interface Serial3/0
#
interface Serial4/0
#
interface NULL0
#
interface GigabitEthernet0/0
 port link-mode route
 combo enable copper
#
interface GigabitEthernet0/1
 port link-mode route
 combo enable copper
 IP address 222.4.10.1 255.255.255.0
#
interface GigabitEthernet0/2
 port link-mode route
 combo enable copper
#
interface GigabitEthernet5/0
 port link-mode route
 combo enable copper
#
interface GigabitEthernet5/1
 port link-mode route
 combo enable copper
#
interface GigabitEthernet6/0
```

```
 port link-mode route
 combo enable copper
#
interface GigabitEthernet6/1
 port link-mode route
 combo enable copper
#
 scheduler logfile size 16
#
line class aux
 user-role network-operator
#
line class console
 user-role network-admin
#
line class tty
 user-role network-operator
#
line class vty
 user-role network-operator
#
line aux 0
 user-role network-operator
#
line con 0
 user-role network-admin
#
line vty 0 63
 user-role network-operator
#
 undo info-center enable
#
domain name system
#
 domain default enable system
#
role name level-0
 descrIPtion Predefined level-0 role
#
role name level-1
 descrIPtion Predefined level-1 role
#
role name level-2
 descrIPtion Predefined level-2 role
#
```

```
role name level-3
 descrIPtion Predefined level-3 role
#
role name level-4
 descrIPtion Predefined level-4 role
#
role name level-5
 descrIPtion Predefined level-5 role
#
role name level-6
 descrIPtion Predefined level-6 role
#
role name level-7
 descrIPtion Predefined level-7 role
#
role name level-8
 descrIPtion Predefined level-8 role
#
role name level-9
 descrIPtion Predefined level-9 role
#
role name level-10
 descrIPtion Predefined level-10 role
#
role name level-11
 descrIPtion Predefined level-11 role
#
role name level-12
 descrIPtion Predefined level-12 role
#
role name level-13
 descrIPtion Predefined level-13 role
#
role name level-14
 descrIPtion Predefined level-14 role
#
user-group system
#
return end
```

查看路由器 R2 的 OSPF 发布的路由及被动接口：

```
[R2]disp cu
#
 version 7.1.075, Alpha 7571
#
```

```
 sysname R2
#
ospf 1
 silent-interface GigabitEthernet0/1
 silent-interface Serial1/0
 area 0.0.0.0
  network 192.168.1.0 0.0.0.255
  network 222.4.0.0 0.0.0.3
#
 system-working-mode standard
 xbar load-single
 password-recovery enable
 lpu-type f-series
#
vlan 1
#
interface Serial1/0
 IP address 222.4.0.2 255.255.255.252
#
interface Serial2/0
#
interface Serial3/0
#
interface Serial4/0
#
interface NULL0
#
interface GigabitEthernet0/0
 port link-mode route
 combo enable copper
#
interface GigabitEthernet0/1
 port link-mode route
 combo enable copper
 IP address 192.168.1.1 255.255.255.0
#
interface GigabitEthernet0/2
 port link-mode route
 combo enable copper
#
interface GigabitEthernet5/0
 port link-mode route
 combo enable copper
#
interface GigabitEthernet5/1
```

```
 port link-mode route
 combo enable copper
#
interface GigabitEthernet6/0
 port link-mode route
 combo enable copper
#
interface GigabitEthernet6/1
 port link-mode route
 combo enable copper
#
 scheduler logfile size 16
#
line class aux
 user-role network-operator
#
line class console
 user-role network-admin
#
line class tty
 user-role network-operator
#
line class vty
 user-role network-operator
#
line aux 0
 user-role network-operator
#
line con 0
 user-role network-admin
#
line vty 0 63
 user-role network-operator
#
 undo info-center enable
#
domain name system
#
 domain default enable system
#
role name level-0
 descrIPtion Predefined level-0 role
#
role name level-1
 descrIPtion Predefined level-1 role
```

```
#
role name level-2
 descrIPtion Predefined level-2 role
#
role name level-3
 descrIPtion Predefined level-3 role
#
role name level-4
 descrIPtion Predefined level-4 role
#
role name level-5
 descrIPtion Predefined level-5 role
#
role name level-6
 descrIPtion Predefined level-6 role
#
role name level-7
 descrIPtion Predefined level-7 role
#
role name level-8
 descrIPtion Predefined level-8 role
#
role name level-9
 descrIPtion Predefined level-9 role
#
role name level-10
 descrIPtion Predefined level-10 role
#
role name level-11
 descrIPtion Predefined level-11 role
#
role name level-12
 descrIPtion Predefined level-12 role
#
role name level-13
 descrIPtion Predefined level-13 role
#
role name level-14
 descrIPtion Predefined level-14 role
#
user-group system
#
return
```

查看路由器 R3 的 OSPF 发布的路由及被动接口：

```
[R3]disp cu
```

```
#
 version 7.1.075, Alpha 7571
#
 sysname R3
#
ospf 1
 silent-interface GigabitEthernet0/1
 area 0.0.0.0
  network 192.168.2.0 0.0.0.255
  network 222.4.0.0 0.0.0.3
#
 system-working-mode standard
 xbar load-single
 password-recovery enable
 lpu-type f-series
#
vlan 1
#
interface Serial1/0
#
interface Serial2/0
 IP address 222.4.0.5 255.255.255.252
#
interface Serial3/0
#
interface Serial4/0
#
interface NULL0
#
interface GigabitEthernet0/0
 port link-mode route
 combo enable copper
#
interface GigabitEthernet0/1
 port link-mode route
 combo enable copper
 IP address 192.168.2.1 255.255.255.0
#
interface GigabitEthernet0/2
 port link-mode route
 combo enable copper
#
interface GigabitEthernet5/0
 port link-mode route
 combo enable copper
```

```
#
interface GigabitEthernet5/1
 port link-mode route
 combo enable copper
#
interface GigabitEthernet6/0
 port link-mode route
 combo enable copper
#
interface GigabitEthernet6/1
 port link-mode route
 combo enable copper
#
 scheduler logfile size 16
#
line class aux
 user-role network-operator
#
line class console
 user-role network-admin
#
line class tty
 user-role network-operator
#
line class vty
 user-role network-operator
#
line aux 0
 user-role network-operator
#
line con 0
 user-role network-admin
#
line vty 0 63
 user-role network-operator
#
 undo info-center enable
#
domain name system
#
domain default enable system
#
role name level-0
 descrIPtion Predefined level-0 role
#
```

```
role name level-1
 descrIPtion Predefined level-1 role
#
role name level-2
 descrIPtion Predefined level-2 role
#
role name level-3
 descrIPtion Predefined level-3 role
#
role name level-4
 descrIPtion Predefined level-4 role
#
role name level-5
 descrIPtion Predefined level-5 role
#
role name level-6
 descrIPtion Predefined level-6 role
#
role name level-7
 descrIPtion Predefined level-7 role
#
role name level-8
 descrIPtion Predefined level-8 role
#
role name level-9
 descrIPtion Predefined level-9 role
#
role name level-10
 descrIPtion Predefined level-10 role
#
role name level-11
 descrIPtion Predefined level-11 role
#
role name level-12
 descrIPtion Predefined level-12 role
#
role name level-13
 descrIPtion Predefined level-13 role
#
role name level-14
 descrIPtion Predefined level-14 role
#
user-group system
#
return
```

④ 整理路由器 R1、R2 和 R3 的 OSPF 配置信息、各接口及主机的 IP 地址规划，如表 9-4 所示。

表 9-4　各路由器的 OSPF 配置信息、各接口及主机的 IP 地址规划

路由器	被动接口	发布路由	接口及主机	IP 地址
R1	无	222.4.10.0 222.4.0.0	G0/1	222.4.10.1
			S1/0	222.4.0.1
			S2/0	222.4.0.6
			Server1	222.4.10.2
R2	S1/0	222.4.0.0 192.168.1.0	S1/0	222.4.0.2
			G0/1	192.168.1.1
			PC0	192.168.1.2
			PC1	192.168.1.3
R3	无	222.4.0.0 192.168.2.0	S2/0	222.4.0.5
			G0/1	192.168.2.1
			PC2	192.168.2.2
			PC3	192.168.2.3

由于主机与各自网关之间可 ping 通，路由器之间也可以 ping 通，因此各主机及接口配置均正确。

2）根据相关知识点确定故障点的位置

分析路由器 R1、R2、R3 的配置信息，发现 R3 的发布路由错误，根据路由协议，R3 的接口应该发布直连接口；又发现 R2 的 S1/0 接口被设置为被动接口，根据路由器被动接口原理，确认 R2 的被动接口 S1/0 是一个故障点。经过分析，最终确认有以下两个问题。

① 路由器 R2 的 S1/0 接口是被动接口。

② 路由器 R3 的发布路由错误。

3）修正配置，并进行测试

① 对于路由器 R2，去掉 S1/0 接口的被动接口配置，命令如下：

```
[R2-ospf-1]ospf 1
[R2-ospf-1]undo silent-interface Serial1/0
```

配置完成后使用 disp cu 命令进行确认。

② 对于路由器 R3，修改 OSPF 发布的路由，命令如下：

```
[R3]ospf 1
[R3-ospf-1]area 0
[R3-ospf-1-area-0.0.0.0]undo net 222.4.0.0 0.0.0.3
[R3-ospf-1-area-0.0.0.0]net 222.4.0.4 0.0.0.3
```

配置完成后使用 disp cu 命令进行确认。

保存修改后的配置信息，并进行终端互联互通测试，结果如表 9-5 所示。

表 9-5　终端互联互通测试结果

测试序号	路由器 R1	路由器 R2	路由器 R3	测试方法	预期测试结果	实际测试结果	是否发生故障
1	Server1	PC0		Web 访问	成功	成功	否
2		PC1 和 R2-F0/0		ping 命令	成功	成功	否
3	Server1		PC2	Web 访问	成功	成功	否
4			PC3 和 R3-F0/0	ping 命令	成功	成功	否
5	R1-S0/0/0	PC1		ping 命令	成功	成功	否
6	R1-S0/0/1		PC3	ping 命令	成功	成功	否
7	Server1 和 R1-F0/0			ping 命令	成功	成功	否
8	R1-S0/0/0	R2-S0/0/0		ping 命令	成功	成功	否
9	R1-S0/0/1		R3-S0/0/0	ping 命令	成功	成功	否

由表 9-5 可知，9 项测试全部成功，由此确认故障现象和以上分析完全吻合，完成故障排除。

4）整理新的配置文档

在故障排除后，保存所有路由器的配置信息，并更新书面的记录材料，确保书面文档和实际配置保持一致，以确保下次配置正常使用。

9.3　相关知识准备

知识准备

为了能够深入地分析故障点，读者应了解 OSPF 的相关知识。

1）什么是 OSPF

OSPF（Open Shortest Path First，开放式最短路径优先）是一个内部网关协议，用于在单一自治系统（Autonomous System，AS）内进行路由决策，具有路由变化收敛速度快、无路由环路、支持变长子网掩码（VLSM）和汇总，以及层次区域划分等优点。与 RIP 相比，OSPF 是链路状态（Link State）路由协议，而 RIP 是距离矢量路由协议。OSPF 的协议管理距离是 110。

OSPF 是一种典型的链路状态路由协议，通常用于单个路由域。在这里，单个路由域是指一个 AS，即一组通过相同的路由政策或路由协议互相交换路由信息的网络。在这个 AS 中，所有 OSPF 路由器都维护一个相同的描述该 AS 结构的数据库。该数据库中存放的是路由域中相应链路的状态信息。OSPF 路由器正是通过这个数据库计算出 OSPF 路由表的。

作为一种链路状态路由协议，OSPF 将 LSA（Link-State Advertisement，链路状态通告）数据包传送给某一区域内的所有路由器，这一点与距离矢量路由协议不同。运行距离矢量路由协议的路由器会将部分或全部路由表传递给与其相邻的路由器。

2）OSPF 特点

① 快速收敛。OSPF 能够在最短时间内将路由变化传递到整个 AS，因此它的收敛速

度快。

② 区域划分。OSPF 引入了区域（Area）划分的概念，在将 AS 划分为不同区域后，通过在区域之间发送路由信息摘要，极大地减少了需要传递的路由信息数量。这种设计使得路由信息不会随着网络规模的扩大而急剧膨胀。

③ 开销控制。将协议自身的开销控制到最小。

3）OSPF 分析

（1）OSPF 数据包。

OSPF 数据包格式如图 9-2 所示。

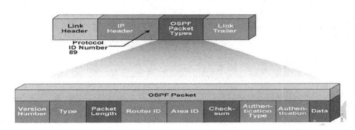

图 9-2　OSPF 数据包格式

在 OSPF 数据包中，其数据包头占 24 字节，包含如下 8 个字段。

① Version Number：用于定义所采用的 OSPF 版本。

② Type：用于定义 OSPF 数据包类型。OSPF 数据包共有如下 5 种类型。

* Hello：用于建立和维护相邻的两个 OSPF 路由器的关系。该数据包是定期发送的。

* Database Description：用于描述整个数据库。该数据包仅在 OSPF 初始化时发送。

* Link state request：用于向相邻的 OSPF 路由器请求部分或全部的数据。这种数据包在路由器发现其数据已经过期时才发送。

* Link state update：对 Link state 请求数据包的响应，即 LSA 数据包。

* Link state acknowledgment：对 LSA 数据包的响应。

③ Packet Length：用于定义整个数据包的长度。

④ Router ID：用于描述数据包的源地址，以 IP 地址来表示。

⑤ Area ID：用于区分 OSPF 数据包所属的区域号。所有的 OSPF 数据包都属于一个特定的 OSPF 区域。

⑥ Checksum：校验位，用于标记数据包在传递时有无误码。

⑦ Authentication Type：用于定义 OSPF 验证类型。

⑧ Authentication：包含 OSPF 验证信息，占 8 字节。

（2）SPF 算法及最短路径树。

SPF 算法是 OSPF 的基础。SPF 算法有时又被称为 Dijkstra 算法。SPF 算法将每台路由器视为根，计算其到每一台目的地路由器的距离。每一台路由器根据一个统一的数据库计算出路由域的网络拓扑结构图。该网络拓扑结构图类似于一棵树，在 SPF 算法中，这棵树被称为最短路径树。在 OSPF 中，最短路径树的树干长度（OSPF 路由器至每一台目的地路由器的距离）被称为 OSPF 的 Cost，其算法为 Cost = $100×10^6$/链路带宽。

在这里，链路带宽以 bps 来表示。也就是说，OSPF 的 Cost 与链路带宽成反比，带宽越高，Cost 越小，表示 OSPF 到目的地的距离越近。例如，FDDI 或快速以太网的 Cost 为 1，2Mbps 串行链路的 Cost 为 48，10Mbps 以太网的 Cost 为 10 等。

（3）链路状态算法。

链路状态算法可以分为以下 4 个步骤。

① 发现邻居节点，并获取它们的地址。

② 当路由器初始化或网络结构发生变化（如增减路由器、链路状态发生变化等）时，路由器会产生 LSA 数据包。该数据包中包含路由器上所有相连链路，即所有接口的状态信息。

③ 所有路由器会通过一种扩散（Flooding）方法来交换链路状态数据。通过扩散，路由器会将其 LSA 数据包传送给所有与其相邻的 OSPF 路由器；相邻路由器根据接收到的链路状态信息更新自己的数据库，并将该链路状态信息传送给相邻的路由器，直至达到稳定状态。

④ 当网络重新稳定时（也可以说当 OSPF 路由协议收敛完成时），所有路由器会根据各自的链路状态信息数据库计算出各自的路由表。该路由表中包含路由器到每一个可到达目的地的 Cost，以及到达该目的地所要转发的下一台路由器信息。

第 4 个步骤实际上是指 OSPF 的一个特性。当网络状态比较稳定时，网络中传递的链路状态信息是比较少的，或者说网络比较"安静"。这也是链路状态路由协议与距离矢量路由协议的一个不同之处。

（4）OSPF 下的被动接口。

① 接口停止发送和接收 hello 数据包，邻居关系将会终止或无法建立。

② 该接口不接收也不发送路由更新信息。

如图 9-3 所示，假设路由器 R1、R2、R3 运行 OSPF，其中 R3 的 G0/0 接口被设置为被动接口，按照 OSPF 下被动接口的特点，将会产生以下结果。

① 路由器 R3 无法与路由器 R1 建立邻居关系。

② 路由器 R3 无法学习任何网段信息。

③ 路由器 R1 无法从路由器 R3 上获取任何网段信息。

④ 路由器 R1、R2 之间正常工作。

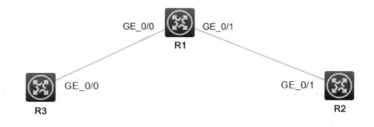

图 9-3　协议配置

（5）引入发布路由。

OSPF 通过 network 命令引入发布路由。此命令具有以下功能。

① 路由器上任何符合 network 命令中的网络地址的接口都将被启用，从而可以发送和接收 OSPF 数据包。

② 此网络（或子网）将被包括在 OSPF 路由更新中。

network 命令在路由器配置模式下的使用形式如下：

```
network network-address wildcard-mask area area-id
```

其中，network-address 为直连网络号，wildcard-mask 为反掩码。

9.4 项目小结

本项目主要针对 OSPF 配置进行故障排除，主要思路是先硬件、后软件。因为本书的项目都是在模拟环境下实现的，所以直接从软件开始进行故障排除。对于本项目，第一，检查 IP 地址规划是否错误；第二，检查是否已经建立 OSPF 邻居关系，并使用命令 display ospf peer brief 查看邻居关系；第三，检查 OSPF 发布的路由是否正确、反掩码是否正确，以及是否已经将关键接口设置为静默（Silent）；第四，检查是否已经配置 OSPF 的接口属于特定的区域，直到检查出故障为止。

素质拓展：移花接木 降本增效

网约车在人们的日常出行生活中扮演了十分重要的角色。它为出租车和乘客搭建了一个方便的业务平台，不仅减少了乘客的候车时间，而且在很大程度上缓解了出租车的空载现象，极大地提高了人们的出行效率。网约车的订单分配模式之一是系统派单模式。在系统派单模式下，订单分配策略旨在保证所有乘客等待的时间全局最短。该模式使用 SPF 算法计算每辆出租车到每位乘客的最短路径，并通过 KM 算法求出租车与乘客之间的最优匹配，以确保所有出租车与乘客之间的行驶距离最短，从而最大程度地减少乘客的候车时间。

怎么样？有趣吧！滴滴打车为了提高客户体验还能借鉴网络动态路由中的最短路径算法。

增值服务

当企业规模较大时，企业网络可以采用 OSPF 进行配置，而随着企业规模不断扩大，企业子网数量增加，会导致网络运行状态不够稳定。这时，售后工程人员可以对企业网络提出优化建议，并且考虑企业的网络安全性和稳定性，对企业网络进行认证接入，同时检测网络运行状态。

认证可以在区域或接口视图下通过 authentication-mode 命令进行设置。检测网络运行状态可以通过 OSPF 与 BFD 联动配置来实现，从而提高网络收敛性和稳定性。

9.5 课后实训

　　项目内容：王某在市区开了一家超市，后来由于业务拓展，又陆续在市中心开了 3 家超市，第一次开的超市为总店，后开的 3 家超市为分店。王某想建设自己的内部局域网，要求使用 OSPF 来实现，并且各个分店能够与总店进行通信，分店与分店之间不允许进行通信。IP 地址的规划如下：总店选用 172.16.1.0/24 网段，网关选用该网段最后一个地址；3 间分店分别选用 172.16.2.0/24、172.16.3.0/24、172.16.4.0/24 网段，总店与各个分店之间的链路地址分别为 172.16.1.0/30、172.16.1.4/30、172.16.1.8/30。

　　在完成网络配置后进行测试时，网络出现了故障现象。请按如下操作步骤排除故障。

① 根据要求检查故障现象。
② 根据故障现象收集故障信息。
③ 利用结构化故障排除方法完成故障定位。
④ 修改故障配置并说明故障原因。
⑤ 更新配置文档。

项目 10

多路由协议共存的故障排除

内容介绍

某公司扩大营销规模并购了两家小公司,由于业务上的需求,需在原有的网络系统上接入新并购的两家子公司的局域网。现公司网络使用的是 OSPF,但新并购的一家公司由于规模小,使用的是静态路由协议,而另一家公司使用的是 RIP。现要求在多路由协议共存的情况下实现业务的互联互通。

任务安排

任务 1　进行多路由协议共存互联的网络配置
任务 2　进行网络配置过程中的故障分析与排除

学习目标

◇ 了解多路由协议共存情况下常见故障的产生原因
◇ 掌握故障排除的思路
◇ 学会结构化故障排除方法
◇ 学会多路由协议共存相关故障排除及文档更新的方法

素质目标

在工作中要独立思考,在实践中要善于创新。

10.1 多路由协议共存配置分析与实施

发现故障

网络管理员王工在接到公司网络接入子公司网络的任务后,根据公司业务需求,完成了物理上的连接,形成了新的网络拓扑结构,如图 10-1 所示。其中,R2 为总公司的外网出口路由器,R2 的 G0/0 接口对应公司总部路由器 R1 的 G0/0 接口,R2 的 G0/1 接口对应子公司路由器 R3 的 G0/0 接口,R2 的 G0/2 接口对应另一子公司路由器 R4 的 G0/0 接口,PC1、PC2 为总公司的内部主机,PC3、PC4 和 PC5、PC6 分别为两家子公司的内部主机。

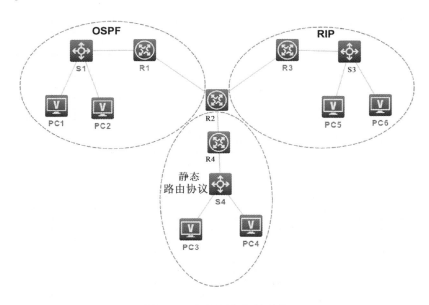

图 10-1 新的网络拓扑结构

王工对设备进行配置后,进行了主机互联互通测试,结果如表 10-1 所示。

表 10-1 主机互联互通测试结果

测试序号	源主机	目的主机	测试方法	预期测试结果	实际测试结果	是否发生故障
1	PC1	PC3	ping 命令	成功	失败	是
2	PC1	PC4	ping 命令	成功	失败	是
3	PC2	PC3	ping 命令	成功	失败	是
4	PC2	PC4	ping 命令	成功	失败	是
5	PC1	PC5	ping 命令	成功	失败	是
6	PC1	PC6	ping 命令	成功	失败	是
7	PC2	PC5	ping 命令	成功	失败	是
8	PC2	PC6	ping 命令	成功	失败	是

由表 10-1 可知,8 项测试全部失败,因此本次操作没有成功完成项目目标。这是什么原因造成的故障现象呢?是规划设计的问题、操作的问题,还是多路由协议共存概念没有

理解清楚的问题？据此，要深入进行故障分析，以确定问题所在。

排除故障

10.2 多路由协议共存配置故障分析与排除

1. 故障分析方法

根据结构化故障排除思路，故障排除有自下而上、自上而下、分而治之及路径排查等方法。由于本项目中的 3 家公司使用的路由协议各不相同，因此采用在每家公司内部分别进行检查的方法，先检查静态路由协议，再检查 RIP，最后检查 OSPF，逐步查找故障原因。

从故障排除理论体系角度进行分析，故障点可能包括以下几方面。

① 静态路由设置错误。

② RIP 版本错误。

③ OSPF 配置故障。

2. 分析故障点

从多路由协议的实现及技术原理进行分析，可能存在以下故障点。

① 路由重分布故障。

② 进程号错误。

③ 发布的区域错误。

由于本次故障主要在模拟环境下实现，因此可以忽略物理问题及设备问题。本次主要从静态路由协议、RIP、OSPF 的配置方面进行故障排除。

本次故障排除任务的目标如下。

① 查看 IP 地址配置，确定已有的配置信息。

② 查看各路由协议配置，确定已有的配置信息。

完成本次故障排除任务所需的设备如下。

① 一台装有超级终端软件或 Telnet 软件的计算机，同时确定访问所需的用户名和口令。

② 配置线缆。

③ 笔和纸，用于记录相关信息。

1）查看现有路由器配置

① 使用超级终端软件或 Telnet 软件连接交换机，并使用 disp ip routing-table 命令查看路由器 R1、R2、R3 和 R4 的路由表：

```
[R1]disp ip routing-table

Destinations : 17     Routes : 17

Destination/Mask   Proto    Pre Cost      NextHop         Interface
0.0.0.0/32         Direct   0   0         127.0.0.1       InLoop0
127.0.0.0/8        Direct   0   0         127.0.0.1       InLoop0
```

```
127.0.0.0/32          Direct    0    0      127.0.0.1       InLoop0
127.0.0.1/32          Direct    0    0      127.0.0.1       InLoop0
127.255.255.255/32    Direct    0    0      127.0.0.1       InLoop0
192.168.10.0/25       Direct    0    0      192.168.10.126  GE0/1
192.168.10.0/32       Direct    0    0      192.168.10.126  GE0/1
192.168.10.126/32     Direct    0    0      127.0.0.1       InLoop0
192.168.10.127/32     Direct    0    0      192.168.10.126  GE0/1
192.168.12.0/30       Direct    0    0      192.168.12.1    GE0/0
192.168.12.0/32       Direct    0    0      192.168.12.1    GE0/0
192.168.12.1/32       Direct    0    0      127.0.0.1       InLoop0
192.168.12.3/32       Direct    0    0      192.168.12.1    GE0/0
192.168.20.0/24       RIP       100  1      192.168.12.2    GE0/0
224.0.0.0/4           Direct    0    0      0.0.0.0         NULL0
224.0.0.0/24          Direct    0    0      0.0.0.0         NULL0
255.255.255.255/32    Direct    0    0      127.0.0.1       InLoop0
[R2]disp ip routing-table

Destinations : 23      Routes : 23

Destination/Mask      Proto     Pre  Cost   NextHop         Interface
0.0.0.0/32            Direct    0    0      127.0.0.1       InLoop0
127.0.0.0/8           Direct    0    0      127.0.0.1       InLoop0
127.0.0.0/32          Direct    0    0      127.0.0.1       InLoop0
127.0.0.1/32          Direct    0    0      127.0.0.1       InLoop0
127.255.255.255/32    Direct    0    0      127.0.0.1       InLoop0
192.168.10.0/24       RIP       100  1      192.168.12.1    GE0/0
192.168.10.128/25     Static    60   0      192.168.24.2    GE0/2
192.168.12.0/30       Direct    0    0      192.168.12.2    GE0/0
192.168.12.0/32       Direct    0    0      192.168.12.2    GE0/0
192.168.12.2/32       Direct    0    0      127.0.0.1       InLoop0
192.168.12.3/32       Direct    0    0      192.168.12.2    GE0/0
192.168.20.0/24       O_INTRA   10   2      192.168.23.2    GE0/1
192.168.23.0/30       Direct    0    0      192.168.23.1    GE0/1
192.168.23.0/32       Direct    0    0      192.168.23.1    GE0/1
192.168.23.1/32       Direct    0    0      127.0.0.1       InLoop0
192.168.23.3/32       Direct    0    0      192.168.23.1    GE0/1
192.168.24.0/30       Direct    0    0      192.168.24.1    GE0/2
192.168.24.0/32       Direct    0    0      192.168.24.1    GE0/2
192.168.24.1/32       Direct    0    0      127.0.0.1       InLoop0
192.168.24.3/32       Direct    0    0      192.168.24.1    GE0/2
224.0.0.0/4           Direct    0    0      0.0.0.0         NULL0
224.0.0.0/24          Direct    0    0      0.0.0.0         NULL0
255.255.255.255/32    Direct    0    0      127.0.0.1       InLoop0
<R3>disp ip routing-table
```

```
Destinations : 17        Routes : 17

Destination/Mask         Proto    Pre  Cost    NextHop           Interface
0.0.0.0/32               Direct   0    0       127.0.0.1         InLoop0
127.0.0.0/8              Direct   0    0       127.0.0.1         InLoop0
127.0.0.0/32             Direct   0    0       127.0.0.1         InLoop0
127.0.0.1/32             Direct   0    0       127.0.0.1         InLoop0
127.255.255.255/32       Direct   0    0       127.0.0.1         InLoop0
192.168.10.0/24          O_ASE2   150  1       192.168.23.1      GE0/0
192.168.20.0/24          Direct   0    0       192.168.20.254    GE0/1
192.168.20.0/32          Direct   0    0       192.168.20.254    GE0/1
192.168.20.254/32        Direct   0    0       127.0.0.1         InLoop0
192.168.20.255/32        Direct   0    0       192.168.20.254    GE0/1
192.168.23.0/30          Direct   0    0       192.168.23.2      GE0/0
192.168.23.0/32          Direct   0    0       192.168.23.2      GE0/0
192.168.23.2/32          Direct   0    0       127.0.0.1         InLoop0
192.168.23.3/32          Direct   0    0       192.168.23.2      GE0/0
224.0.0.0/4              Direct   0    0       0.0.0.0           NULL0
224.0.0.0/24             Direct   0    0       0.0.0.0           NULL0
255.255.255.255/32       Direct   0    0       127.0.0.1         InLoop0
<R4>disp ip routing-table

Destinations : 17        Routes : 17

Destination/Mask         Proto    Pre  Cost    NextHop           Interface
0.0.0.0/0                Static   60   0       192.168.24.1      GE0/0
0.0.0.0/32               Direct   0    0       127.0.0.1         InLoop0
127.0.0.0/8              Direct   0    0       127.0.0.1         InLoop0
127.0.0.0/32             Direct   0    0       127.0.0.1         InLoop0
127.0.0.1/32             Direct   0    0       127.0.0.1         InLoop0
127.255.255.255/32       Direct   0    0       127.0.0.1         InLoop0
192.168.10.128/25        Direct   0    0       192.168.10.254    GE0/1
192.168.10.128/32        Direct   0    0       192.168.10.254    GE0/1
192.168.10.254/32        Direct   0    0       127.0.0.1         InLoop0
192.168.10.255/32        Direct   0    0       192.168.10.254    GE0/1
192.168.24.0/30          Direct   0    0       192.168.24.2      GE0/0
192.168.24.0/32          Direct   0    0       192.168.24.2      GE0/0
192.168.24.2/32          Direct   0    0       127.0.0.1         InLoop0
192.168.24.3/32          Direct   0    0       192.168.24.2      GE0/0
224.0.0.0/4              Direct   0    0       0.0.0.0           NULL0
224.0.0.0/24             Direct   0    0       0.0.0.0           NULL0
255.255.255.255/32       Direct   0    0       127.0.0.1         InLoop0
```

② 使用 Telnet 软件查看路由器各接口及主机的 IP 地址配置，如表 10-2、表 10-3 所示。

表 10-2 路由器各接口的 IP 地址配置

本地设备	接口	IP 地址	对端设备	接口	IP 地址	掩码
R1	G0/0	192.168.12.1	R2	G0/0	192.168.12.2	255.255.255.252
R2	G0/1	192.168.23.1	R3	G0/0	192.168.23.2	255.255.255.252
R2	G0/2	192.168.24.1	R4	G0/0	192.168.24.2	255.255.255.252

表 10-3 主机的 IP 地址配置

主机	IP 地址	掩码	网关
PC1	192.168.10.1	255.255.255.128	192.168.10.126
PC2	192.168.10.2	255.255.255.128	192.168.10.126
PC3	192.168.10.129	255.255.255.128	192.168.10.254
PC4	192.168.10.130	255.255.255.128	192.168.10.254
PC5	192.168.20.1	255.255.255.0	192.168.20.254
PC6	192.168.20.2	255.255.255.0	192.168.20.254

通过对路由器各接口及主机 IP 地址进行核对，确认各个终端的 IP 地址没有错误，现有 IP 地址配置设计合理，但发现路由器 R1 的路由表中不存在新增的两家子公司的网络条目。

2）根据相关知识点确定故障点的位置

根据上面的分析，故障点主要是公司总部的路由器 R1 不能从 R2 上学习到新加入的两个家公司的路由信息，因此从此处着手开始排除故障。

使用 disp ip routing-table protocol rip 命令分别查看路由器 R1 与 R2 的路由协议的主要配置信息：

```
[R1]disp ip routing-table protocol rip
Summary count : 3
RIP Routing table status : <Active>
Summary count : 1
Destination/Mask      Proto   Pre Cost        NextHop            Interface
192.168.20.0/24       RIP     100 1           192.168.12.2       GE0/0
RIP Routing table status : <Inactive>
Summary count : 2
Destination/Mask      Proto   Pre Cost        NextHop            Interface
192.168.10.0/25       RIP     100 0           0.0.0.0            GE0/1
192.168.12.0/30       RIP     100 0           0.0.0.0            GE0/0
[R2]disp ip routing-table protocol rip
Summary count : 2
RIP Routing table status : <Active>
Summary count : 1
Destination/Mask      Proto   Pre Cost        NextHop            Interface
192.168.10.0/24       RIP     100 1           192.168.12.1       GE0/0
RIP Routing table status : <Inactive>
Summary count : 1
Destination/Mask      Proto   Pre Cost        NextHop            Interface
```

```
192.168.12.0/30      RIP       100 0       0.0.0.0           GE0/0
[R2]disp ip routing-table protocol ospf
Summary count : 2
OSPF Routing table status : <Active>
Summary count : 1
Destination/Mask    Proto     Pre Cost    NextHop           Interface
192.168.20.0/24     O_INTRA   10  2       192.168.23.2      GE0/1

OSPF Routing table status : <Inactive>
Summary count : 1

Destination/Mask    Proto     Pre Cost    NextHop           Interface
192.168.23.0/30     O_INTRA   10  1       0.0.0.0           GE0/1
[R2]disp ip routing-table protocol static
Summary count : 1
Static Routing table status : <Active>
Summary count : 1

Destination/Mask    Proto     Pre Cost    NextHop           Interface
192.168.10.128/25   Static    60  0       192.168.24.2      GE0/2
Static Routing table status : <Inactive>
Summary count : 0
```

分析上面的信息可知，路由器 R1 与 R2 交互的路由协议为 RIP，R2 上同时存在 RIP、OSPF 和静态路由协议，并且 R2 已经将 OSPF 和静态路由协议都重分布到 RIP 中，但是在 R1 上为什么学习不到重分布的路由呢？进一步分析上面的信息，发现 R1 和 R2 使用的协议都是 RIPv1。根据 RIPv1 的特征，RIPv1 是有类路由协议，它在路由更新中不带子网掩码，在同一网络的有类边界会自动汇总，而 PC1 和 PC2 所在的网段（192.168.10.0/25）和重分布到 R2 的 RIP 中的静态路由（PC3 和 PC4 所在的网段 192.168.10.128/25）属于同一 C 类地址段 192.168.10.0/24。由此可以确定本项目的故障点之一是 R1 和 R2 使用的 RIP 版本不同。

3）修改配置，并进行测试

分别修改路由器 R1 和 R2 的 RIP 版本号：

```
[R1]rip 1
[R1-rip-1]vers 2
[R1-rip-1]undo summary
[R2]rip 1
[R2-rip-1]vers 2
[R2-rip-1]undo summary
```

在修改配置后，查看路由表：

```
[R1]disp ip routing-table

Destinations : 17      Routes : 17
```

Destination/Mask	Proto	Pre	Cost	NextHop	Interface
0.0.0.0/32	Direct	0	0	127.0.0.1	InLoop0
127.0.0.0/8	Direct	0	0	127.0.0.1	InLoop0
127.0.0.0/32	Direct	0	0	127.0.0.1	InLoop0
127.0.0.1/32	Direct	0	0	127.0.0.1	InLoop0
127.255.255.255/32	Direct	0	0	127.0.0.1	InLoop0
192.168.10.0/25	Direct	0	0	192.168.10.126	GE0/1
192.168.10.0/32	Direct	0	0	192.168.10.126	GE0/1
192.168.10.126/32	Direct	0	0	127.0.0.1	InLoop0
192.168.10.127/32	Direct	0	0	192.168.10.126	GE0/1
192.168.10.128/25	RIP	100	1	192.168.12.2	GE0/0
192.168.12.0/30	Direct	0	0	192.168.12.1	GE0/0
192.168.12.0/32	Direct	0	0	192.168.12.1	GE0/0
192.168.12.1/32	Direct	0	0	127.0.0.1	InLoop0
192.168.12.3/32	Direct	0	0	192.168.12.1	GE0/0
224.0.0.0/4	Direct	0	0	0.0.0.0	NULL0
224.0.0.0/24	Direct	0	0	0.0.0.0	NULL0
255.255.255.255/32	Direct	0	0	127.0.0.1	InLoop0

```
[R2]disp ip routing-table
Destinations : 22    Routes : 22
```

Destination/Mask	Proto	Pre	Cost	NextHop	Interface
0.0.0.0/32	Direct	0	0	127.0.0.1	InLoop0
127.0.0.0/8	Direct	0	0	127.0.0.1	InLoop0
127.0.0.0/32	Direct	0	0	127.0.0.1	InLoop0
127.0.0.1/32	Direct	0	0	127.0.0.1	InLoop0
127.255.255.255/32	Direct	0	0	127.0.0.1	InLoop0
192.168.10.128/25	Static	60	0	192.168.24.2	GE0/2
192.168.12.0/30	Direct	0	0	192.168.12.2	GE0/0
192.168.12.0/32	Direct	0	0	192.168.12.2	GE0/0
192.168.12.2/32	Direct	0	0	127.0.0.1	InLoop0
192.168.12.3/32	Direct	0	0	192.168.12.2	GE0/0
192.168.20.0/24	O_INTRA	10	2	192.168.23.2	GE0/1
192.168.23.0/30	Direct	0	0	192.168.23.1	GE0/1
192.168.23.0/32	Direct	0	0	192.168.23.1	GE0/1
192.168.23.1/32	Direct	0	0	127.0.0.1	InLoop0
192.168.23.3/32	Direct	0	0	192.168.23.1	GE0/1
192.168.24.0/30	Direct	0	0	192.168.24.1	GE0/2
192.168.24.0/32	Direct	0	0	192.168.24.1	GE0/2
192.168.24.1/32	Direct	0	0	127.0.0.1	InLoop0
192.168.24.3/32	Direct	0	0	192.168.24.1	GE0/2
224.0.0.0/4	Direct	0	0	0.0.0.0	NULL0
224.0.0.0/24	Direct	0	0	0.0.0.0	NULL0
255.255.255.255/32	Direct	0	0	127.0.0.1	InLoop0

此时，重新进行主机互联互通测试，结果如表10-4所示。

表 10-4　主机互联互通测试结果（1）

测试序号	源主机	目的主机	测试方法	预期测试结果	实际测试结果	是否发生故障
1	PC1	PC3	ping 命令	成功	成功	否
2	PC1	PC4	ping 命令	成功	成功	否
3	PC2	PC3	ping 命令	成功	成功	否
4	PC2	PC4	ping 命令	成功	成功	否
5	PC1	PC5	ping 命令	成功	失败	是
6	PC1	PC6	ping 命令	成功	失败	是
7	PC2	PC5	ping 命令	成功	失败	是
8	PC2	PC6	ping 命令	成功	失败	是

在进行了一系列的故障排除后，已经解决 PC1、PC2 和 PC3、PC4 之间的互联互通问题。但是，PC1、PC2 和 PC5、PC6 之间仍然存在故障，主要原因还是路由器 R2 没有将从 OSPF 重分布到 RIP 的路由信息发送给 R1。为处理上述故障，需要进一步查看路由器 R2 的配置信息。

使用 disp cu 命令查看路由器 R2 的配置信息：

```
[R2]disp cu
#
 version 7.1.075, Alpha 7571
#
 sysname R2
#
ospf 1                                  //路由协议为 OSPF
 import-route rip 1                     //引入 RIP
 area 0.0.0.0
  network 192.168.23.0 0.0.0.3
#
rip 1                                   //启用 RIP
 version 2
 network 192.168.12.0
 import-route static                    //引入静态路由
 import-route ospf 1                    //引入 OSPF
#
 system-working-mode standard
 xbar load-single
 password-recovery enable
 lpu-type f-series
#
vlan 1
#
interface Serial1/0
#
interface Serial2/0
#
```

```
interface Serial3/0
#
interface Serial4/0
#
interface NULL0
#
interface GigabitEthernet0/0
 port link-mode route
 combo enable copper
 ip address 192.168.12.2 255.255.255.252
#
interface GigabitEthernet0/1
 port link-mode route
 combo enable copper
 ip address 192.168.23.1 255.255.255.252
#
interface GigabitEthernet0/2
 port link-mode route
 combo enable copper
 ip address 192.168.24.1 255.255.255.252
#
interface GigabitEthernet5/0
 port link-mode route
 combo enable copper
#
interface GigabitEthernet5/1
 port link-mode route
 combo enable copper
#
interface GigabitEthernet6/0
 port link-mode route
 combo enable copper
#
interface GigabitEthernet6/1
 port link-mode route
 combo enable copper
#
 scheduler logfile size 16
#
line class aux
 user-role network-operator
#
……

line con 0
```

```
 user-role network-admin
#
line vty 0 63
 user-role network-operator
#
 ip route-static 192.168.10.128 25 192.168.24.2
#
 undo info-center enable
#
domain name system
#
domain default enable system
#
role name level-0
 description Predefined level-0 role
#
......
role name level-14
 description Predefined level-14 role
#
user-group system
#
return
```

结合 RIP 的原理分析上面的运行配置文档，发现在将 OSPF 重分布到 RIP 中的命令中没有设置度量值，这可能是导致路由器 R1 不能从 R2 上学习到重分布到 RIP 中的 OSPF 的路由条目的原因。因为 RIP 以跳数为度量值，最大度量值为 15，16 表示不可达，所以一般在将别的动态路由协议重分布到 RIP 中时必须注意度量值的设置。

修改路由器 R2 的配置：

```
[R2]rip 1
[R2-rip-1]import-route ospf 1 cost 5
```

查看路由器 R1 的路由表：

```
[R1]disp ip routing-table

Destinations : 18      Routes : 18

Destination/Mask        Proto    Pre  Cost    NextHop          Interface
0.0.0.0/32              Direct   0    0       127.0.0.1        InLoop0
127.0.0.0/8             Direct   0    0       127.0.0.1        InLoop0
127.0.0.0/32            Direct   0    0       127.0.0.1        InLoop0
127.0.0.1/32            Direct   0    0       127.0.0.1        InLoop0
127.255.255.255/32      Direct   0    0       127.0.0.1        InLoop0
192.168.10.0/25         Direct   0    0       192.168.10.126   GE0/1
192.168.10.0/32         Direct   0    0       192.168.10.126   GE0/1
192.168.10.126/32       Direct   0    0       127.0.0.1        InLoop0
```

```
192.168.10.127/32      Direct  0    0    192.168.10.126    GE0/1
192.168.10.128/25      RIP     100  1    192.168.12.2      GE0/0
192.168.12.0/30        Direct  0    0    192.168.12.1      GE0/0
192.168.12.0/32        Direct  0    0    192.168.12.1      GE0/0
192.168.12.1/32        Direct  0    0    127.0.0.1         InLoop0
192.168.12.3/32        Direct  0    0    192.168.12.1      GE0/0
192.168.20.0/24        RIP     100  6    192.168.12.2      GE0/0
224.0.0.0/4            Direct  0    0    0.0.0.0           NULL0
224.0.0.0/24           Direct  0    0    0.0.0.0           NULL0
255.255.255.255/32     Direct  0    0    127.0.0.1         InLoop0
```

至此，路由器 R1 的路由表学习到了新增的两家子公司的网络条目。

4）在新配置环境下进行测试

在完成配置后，重新对主机进行互联互通测试，结果如表 10-5 所示。

表 10-5 主机互联互通测试结果（2）

测试序号	源主机	目的主机	测试方法	预期测试结果	实际测试结果	是否发生故障
1	PC1	PC3	ping 命令	成功	成功	否
2	PC1	PC4	ping 命令	成功	成功	否
3	PC2	PC3	ping 命令	成功	成功	否
4	PC2	PC4	ping 命令	成功	成功	否
5	PC1	PC5	ping 命令	成功	成功	否
6	PC1	PC6	ping 命令	成功	成功	否
7	PC2	PC5	ping 命令	成功	成功	否
8	PC2	PC6	ping 命令	成功	成功	否

由表 10-5 可知，8 项测试全部成功，由此确定故障现象和以上分析完全吻合。在经过故障点分析和故障点确定，以及分步骤修改配置后，成功完成故障排除工作。

5）整理新的配置文档

在故障排除后，保存所有路由器的配置信息，并更新书面的记录材料，确保书面文档和实际配置保持一致，以确保下次配置正常使用。

10.3 相关知识准备

知识准备

为了能够深入地分析故障点，读者应了解多路由协议的路由引入相关知识。

在进行网络设计时，一般都仅选择运行一种路由协议，以降低网络的复杂度，使网络易于维护。但是在现实中，当需要更换路由协议或对运行不同路由协议的网络进行合并时，有可能在网络中同时运行多路由协议。本项目介绍的便是在多路由协议网络运行环境下，如何进行路由协议之间的引入和部署。

1）多路由协议网络

如果一个网络同时运行了两种及以上路由协议，如同时运行了 OSPF 和 RIP，或者同时运行了动态路由协议和静态路由协议，那么这个网络就是多协议网络。

路由器维护了一张路由表，其中包含来自不同路由协议的路由。因为不同路由协议之间的算法和度量值不同，所以不同路由协议学习到的路由信息不能直接互通，一个路由协议学习到的路由信息不能直接传送到另一个路由协议中。

在对网络进行合并、升级和迁移的过程中，经常会出现多路由协议共存的情况。例如，早期网络中使用了 RIP，但随着网络规模的扩大，当路由器的数量超过 15 台时，RIP 就变得不再适用了。此时，网络管理员可以将 RIP 升级成 OSPF，但在升级过程中可能会出现两种路由协议共同运行的情况。又如，两家公司网络运行了不同的路由协议，当合并这两家公司的网络时，会出现两种路由协议共同运行的情况。

当网络中运行多种路由协议时，需要使用路由引入来将一种路由协议的路由信息引入另一种路由协议，以实现网络互通的目的。

在如图 10-2 所示的路由引入网络拓扑结构中，RTA 和 RTB 运行 OSPF，RTB 和 RTC 运行 RIP；RTA 连接到网络 172.0.0.0/16，RTC 连接到网络 10.0.0.0/24。因为 RTA 和 RTC 运行不同的路由协议，所以它们无法相互学习路由信息，也无法互通。然而，RTB 既运行了 OSPF，又运行了 RIP，因此能够学习到网络 172.0.0.0/16 和 10.0.0.0/24，可以通过路由引入使 RIP 和 OSPF 相互学习到对方的路由信息。

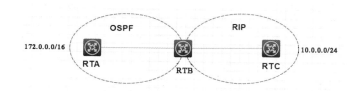

图 10-2　路由引入网络拓扑结构

2）路由引入

通过使用路由引入，网络管理员可以将路由信息从一种路由协议导入另一种路由协议，或者在同种路由协议的不同进程之间导入路由信息。

路由引入通常在边界路由器上进行。边界路由器是同时运行两种及以上路由协议的路由器，它作为不同路由协议之间的桥梁，负责不同路由协议之间的路由引入操作。

在如图 10-2 所示的路由引入网络拓扑结构中，RTB 作为边界路由器，同时运行 OSPF 和 RIP。它一方面与 RTA 通过 OSPF 交换路由信息，另一方面与 RTC 通过 RIP 交换路由信息。在 RTB 上实施路由引入后，RTB 先将通过 RIP 学习到的路由导入 OSPF 的链路状态数据库（Link State DataBase，LSDB），然后以 LSA 数据包的形式将该路由发送到 RTA 上。这样，RTA 的路由表中就有了 10.0.0.0/24 这条路由。同理，RTB 将 OSPF 路由引入 RIP 路由表后，RTC 就学习到了 172.0.0.0/16 这条路由。

注意：只有协议路由表中的有效路由才能成功被引入。

在进行路由引入时，因为不同协议的路由属性表达方式不同，所以原路由属性会发生变化；又因为不同协议的度量值算法不同，所以在进行路由引入时，无法同时引入路由信息的原度量值。此时，协议一般会赋予路由信息一个新的默认度量值（又被称为种子度量值），路由信息在路由器之间传播时，会以新的默认度量值为基础进行度量值的计算。默认度量值可以根据网络的实际情况进行设置，通常被设定为大于路由域内已有路由信息的最大度量值。这样设置可以表示这是从域外引入的路由，以避免潜在的次优路由问题。

有些路由协议会对引入的路由赋予特殊的标记，以表明此路由是从其他路由协议中引入的。例如，OSPF 会把所有引入的外部路由标记为"第二类外部路由（Type2 External）"，并赋予其一个路由标记值 1；IS-IS 协议会把引入的路由放到 Level1-2 路由表中，并设定外部路由开销为 0。

3）路由引入规划

在网络中运行多路由协议会增加网络的复杂度。不同路由协议的算法不同，路由属性不同，收敛速度也不同，混合使用可能会造成次优路由或路由收敛不一致的情况。运行多路由协议对路由器的 CPU、内存等资源有较高要求，所以只在必要时才运行多路由协议。

常见的运行多路由协议的场景有以下 3 种。

① 在对网络进行升级、合并和迁移时出现多协议共存的情况。此时，一般会采用先让两种路由协议共存，再逐步切换到新路由协议的方法。在路由协议共存期间，会使用路由引入来使两种路由协议相互学习彼此的路由信息。

② 在网络中，不是所有设备都支持相同的路由协议。小的接入层设备可能不支持复杂的路由协议，或者某个厂家的设备运行自己的私有协议。在这种情况下，可以规划一部分设备运行一种路由协议，另一部分设备运行另一种路由协议，随后在边界路由器上实施路由引入。

③ 在不同的路由域之间进行路由控制。因为不同路由协议之间不能自动学习路由，所以可以在网络中使用多路由协议，以划分出不同的路由域。在域的边界处，可以进行路由引入时的路由控制。

在多路由协议网络规划中，通常在核心网络中运行链路状态路由协议，如 OSPF、IS-IS 等，以加快收敛速度，提高网络的可靠性；在边缘网络中运行简单的路由协议，如 RIP 或静态路由协议。此时，在实施路由引入时，通常会把路由从边缘网络引入核心网络，并在边缘网络配置静态路由指向核心网络。如果网络中同时运行 IGP 和 BGP，则通常先把 IGP 引入 BGP，再通过 BGP 与外界网络交换路由，以利用 BGP 丰富的属性进行路由控制与选路。

在实施路由引入时，可以仅在一台边界路由器上引入，这被称为单边界引入；也可以在多台边界路由器上引入，这被称为多边界引入。在单边界引入时，相当于两个路由域之间仅有一个连接点，可靠性相对较差。在多边界引入时，不同路由域之间有多条路径，可以提高可靠性，但是这会增加配置的复杂度，并增加产生次优路由的可能性。

一般，应在边缘网络路由器上配置静态或默认路由，下一跳指向核心网络的边界路由器；同时，可以由核心网络的边界路由器发布默认路由。单向路由引入适用于星型拓扑网络。

在边界路由器上把两个路由域的路由相互引入，这被称为双向路由引入。在如图 10-2 所示的路由引入网络拓扑结构中，边界路由器 RTB 把 OSPF 路由域中的路由 172.0.0.0/16

引入 RIP 路由域，同时把 RIP 路由域中的路由 10.0.0.0/24 引入 OSPF 路由域。这样，RTA 和 RTC 就可以知道彼此的具体路由。

在路由引入过程中，系统会只引入路由表中的有效路由，并且本地路由表中不会出现引入后的路由，而是将该路由传递给其他路由器。

路由引入命令请查看相关 H3C 设备手册，这里不再描述。

10.4 项目小结

本项目主要对多路由协议共存所引发的故障进行排除，解决多路由协议共存中的 IP 地址、路由重发布、多协议版本、度量值等一系列问题故障。

素质拓展：雪人计划 利国利民

"雪人计划"是基于全新技术架构的全球下一代互联网（IPv6，互联网协议第六版）根服务器测试和运营实验项目，旨在为下一代互联网提供更多的根服务器解决方案。

"雪人计划"由中国下一代互联网工程中心领衔发起，已在全球 16 个国家完成 25 台 IPv6 根服务器的架设，其中在中国部署了 4 台（由 1 台主根服务器和 3 台辅根服务器组成），打破了中国过去没有根服务器的困境。

增值服务

在多路由协议并存的情况下，需要执行路由引入功能，以实现全网的互联互通。但是，路由引入容易产生次优路由和路由环路的问题。售后工程人员在服务过程中，需要关注企业网络的路由质量和稳定性。在通常情况下，可以利用路由过滤、调整路由协议优先级等方式来避免出现这类问题，从而实现全网互通。

10.5 课后实训

项目内容：X 公司在收购 W 公司后，由于业务需求，需要将两家公司现有的网络进行合并。目前，X 公司网络内部使用的路由协议为 OSPF，网段为 10.10.1.0/24，W 公司网络内部使用的路由协议为 RIP，网段为 172.1.6.1.0/24。现在需要将 W 公司的路由引入 X 公司，将 X 公司的路由引入 W 公司，实现 X 公司的终端与 W 公司的终端的互联互通。

在完成网络配置后进行测试时，网络出现了故障现象。请结合所学知识点，按如下操作步骤排除故障。

① 根据要求检查故障现象。
② 根据故障现象收集故障信息。
③ 利用结构化故障排除方法完成故障定位。
④ 修改故障配置并说明故障原因。
⑤ 更新配置文档。

项目 11

NAT 网络应用的故障排除

内容介绍

某企业向当地的 ISP 供应商申请了 6 个公网 IP 地址，地址范围是 202.20.1.1～202.20.1.6。其中，一个地址分配给 WWW 服务器，以便外网用户可以通过该 IP 地址访问企业官网，了解企业的文化、发展思路及方向等。另外，为了保障企业内部的网络安全，需要提供一个一致的内部网络编址方案，将另外一个地址分配给企业内网的出口，让领导管理人员能够访问外网资料。将剩余的 4 个地址配置为地址池，以供企业内部员工访问外网。

任务安排

任务 1　针对业务需求实现 NAT 网络配置
任务 2　进行网络更新过程中的故障分析与排除

学习目标

◇ 了解 NAT 常见故障的产生原因
◇ 掌握故障排除的思路
◇ 学会结构化故障排除方法
◇ 学会 NAT 相关故障排除及文档更新的方法

素质目标

感受并紧跟科技的发展，增强民族自豪感和自信心，勇于运用网络技术解决现实问题。

11.1 NAT 配置分析与实施

发现故障

网络管理员李工在接到企业网络 NAT 更新任务后，根据企业要求，制定了网络实施方案，主要操作如下。

① 查看现有交换机、路由器的配置，了解现有网络的拓扑结构。
② 将申请到的 IP 地址应用到出口路由器上。
③ 对出口路由器进行配置。
④ 进行 NAT 的互通测试，并分析测试过程中数据包的转发过程。

企业网络拓扑结构如图 11-1 所示。其中，路由器 R1 的 GE_5/0 接口连接 WWW 服务器，GE_0/0 接口连接内网边界路由器 R2，GE_0/1 接口连接交换机 S1，GE_0/2 接口连接交换机 S2；在边界路由器 R2 中，GE_0/1 接口回接内网路由器 R1，GE_0/0 接口的 IP 地址为 ISP 提供的 IP 地址，并连接 Internet；在领导管理人员办公网中，接入交换机 S1 的 GE_0/1 接口连接 PC1，GE_0/2 接口连接 PC2；在员工办公网中，接入交换机 S2 的 GE_0/1 接口连接 PC3，GE_0/2 接口连接 PC4。将 IP 地址 202.20.1.1 应用到 WWW 服务器，通过这个 IP 地址将 WWW 服务器映射到外网，以供外网用户访问；领导管理人员办公网通过边界路由器 R2 的 GE_0/0 接口的 IP 地址（202.20.1.6）来访问 Internet 资源；由于员工办公网中的用户比较多，外网访问需求量大，因此特提供 NAT 地址池（IP 地址范围为 202.20.1.2～202.20.1.5）供用户访问 Internet 资源。

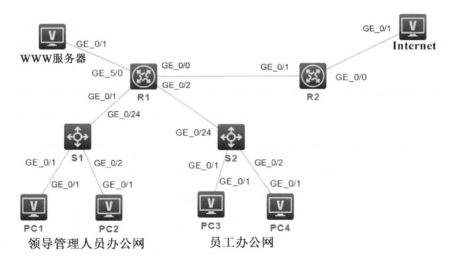

图 11-1　企业网络拓扑结构

在完成配置后，李工进行了终端互联互通测试，结果如表 11-1 所示。由表 11-1 可知，有 4 项测试失败，因此本次操作没有成功完成项目目标。这是什么原因造成的故障现象呢？是规划设计的问题、操作的问题，还是 NAT 技术没有理解清楚的问题？据此，要深入进行故障分析，以确定问题所在。

表 11-1 终端互联互通测试结果

测试序号	本端设备	对端设备	测试方法	预期测试结果	实际测试结果	是否发生故障
1	PC1	Server 服务器	ping 命令	成功	失败	是
2	PC2	Server 服务器	ping 命令	成功	失败	是
3	PC3	Server 服务器	ping 命令	成功	失败	是
4	PC4	Server 服务器	ping 命令	成功	失败	是

11.2 NAT 配置故障分析与排除

排除故障

1. 故障分析方法

根据结构化故障排除思路，严格执行故障排除的操作步骤。首先，确定故障现象并进行详细记录。其次，收集设备信息。本项目主要收集交换机的配置信息、IP 地址、接口信息息、NAT 信息等。再次，在收集信息后，结合 NAT 的实现原理进行综合分析，确定并罗列可能存在的故障点。最后，针对故障点分析出最有可能的故障原因，并对这个原因进行故障排除。本次故障排除可以分解为以下几个任务。

① 查看所有路由器的配置，并按照要求记录在表格中。
② 查看路由器 R2 的 NAT 配置。
③ 根据相关知识点确定故障点的位置。

2. 分析故障点

从 NAT 的实现及技术原理进行分析，可能存在以下故障点。

① 是否设置了访问控制列表（ACL），从而阻塞了网络流量。在进行配置时，要牢记与 ACL 相关的 NAT 操作。如果配置了 ACL 以阻止未进行 NAT 的流量，但实际到达的流量已经进行了地址转换，则可能导致流量被丢弃。

② 在定义要进行 NAT 的 ACL 中，遗漏了需要进行地址转换的网络。ACL 用来定义需要进行 NAT 操作的网络地址，应该包括所有需要进行 NAT 的网络。如果 ACL 中缺少一个或多个地址，那么将导致无法对这些地址的流量进行 NAT。

③ 在 NAT 语句中使用了 no-pat 关键字。为了建立 pat，在 NAT 配置命令的最后不允许使用 no-pat 关键字。

④ 不对称路由导致 NAT 失败。当分组进入一个使用 ip nat inside 命令进行配置的接口，而从另一个使用 ip nat outside 命令进行配置的接口离开时，必须确保需要进行 NAT 的流量进入路由器的所有接口都使用 ip nat inside 命令进行配置，而这个流量离开的所有接口都使用 ip nat outside 命令进行配置，否则流量在经过没有使用正确的 NAT 命令进行配置的接口时无法进行 NAT。

3. 确认故障所需设备

① 一台装有超级终端软件或 Telnet 软件的计算机，同时确定访问所需的用户名和口令。
② 配置线缆。
③ 笔和纸，用于记录相关信息。

4. 实施步骤

1）查看所有路由器的配置

① 使用超级终端软件或 Telnet 软件连接路由器 R1、R2，并使用 disp cu 命令查看路由器接口的 IP 地址和静态路由的配置：

```
[R1]disp cu
#
 version 7.1.075, Alpha 7571
#
 sysname R1
#
 system-working-mode standard
 xbar load-single
 password-recovery enable
 lpu-type f-series
#
vlan 1
#
interface Serial1/0
#
interface Serial2/0
#
interface Serial3/0
#
interface Serial4/0
#
interface NULL0
#
interface GigabitEthernet0/0
 port link-mode route
 combo enable copper
 ip address 172.16.10.1 255.255.255.0
#
interface GigabitEthernet0/1
 port link-mode route
 combo enable copper
 ip address 172.16.1.254 255.255.255.0
#
interface GigabitEthernet0/2
```

```
 port link-mode route
 combo enable copper
 ip address 172.16.2.254 255.255.255.0
#
interface GigabitEthernet5/0
 port link-mode route
 combo enable copper
 ip address 172.16.0.254 255.255.255.0
#
interface GigabitEthernet5/1
 port link-mode route
 combo enable copper
#
interface GigabitEthernet6/0
 port link-mode route
 combo enable copper
#
interface GigabitEthernet6/1
 port link-mode route
 combo enable copper
#
 scheduler logfile size 16
#
line class aux
 user-role network-operator
#
line class console
 user-role network-admin
#
line class tty
 user-role network-operator
#
line class vty
 user-role network-operator
#
line aux 0
 user-role network-operator
#
line con 0
 user-role network-admin
#
line vty 0 63
 user-role network-operator
#
 ip route-static 202.20.1.0 24 GigabitEthernet0/0 172.16.10.2
```

```
#
 undo info-center enable
#
domain name system
#
 domain default enable system
#
role name level-0
 description Predefined level-0 role
#
role name level-1
 description Predefined level-1 role
#
role name level-2
 description Predefined level-2 role
#
role name level-3
 description Predefined level-3 role
#
role name level-4
 description Predefined level-4 role
#
role name level-5
 description Predefined level-5 role
#
role name level-6
 description Predefined level-6 role
#
role name level-7
 description Predefined level-7 role
#
role name level-8
 description Predefined level-8 role
#
role name level-9
 description Predefined level-9 role
#
role name level-10
 description Predefined level-10 role
#
role name level-11
 description Predefined level-11 role
#
role name level-12
 description Predefined level-12 role
```

```
#
role name level-13
 description Predefined level-13 role
#
role name level-14
 description Predefined level-14 role
#
user-group system
#
return
[R2]disp cu
#
 version 7.1.075, Alpha 7571
#
 sysname R2
#
 system-working-mode standard
 xbar load-single
 password-recovery enable
 lpu-type f-series
#
vlan 1
#
interface Serial1/0
#
interface Serial2/0
#
interface Serial3/0
#
interface Serial4/0
#
interface NULL0
#
interface GigabitEthernet0/0
 port link-mode route
 combo enable copper
 ip address 202.20.1.1 255.255.255.0
 nat server protocol tcp global 202.20.1.1 80 inside 172.16.0.1 80
#
interface GigabitEthernet0/1
 port link-mode route
 combo enable copper
 ip address 172.16.10.2 255.255.255.0
#
interface GigabitEthernet0/2
```

```
 port link-mode route
 combo enable copper
#
interface GigabitEthernet5/0
 port link-mode route
 combo enable copper
#
interface GigabitEthernet5/1
 port link-mode route
 combo enable copper
#
interface GigabitEthernet6/0
 port link-mode route
 combo enable copper
#
interface GigabitEthernet6/1
 port link-mode route
 combo enable copper
#
 scheduler logfile size 16
#
line class aux
 user-role network-operator
#
line class console
 user-role network-admin
#
line class tty
 user-role network-operator
#
line class vty
 user-role network-operator
#
line aux 0
 user-role network-operator
#
line con 0
 user-role network-admin
#
line vty 0 63
 user-role network-operator
#
 ip route-static 172.16.0.0 16 172.16.10.1
#
 undo info-center enable
```

```
#
acl basic 2000
 rule 0 permit source 172.16.1.0 0.0.0.255
#
acl basic 2001
 rule 0 permit source 172.16.2.0 0.0.0.255
#
domain name system
#
 domain default enable system
#
role name level-0
 description Predefined level-0 role
#
role name level-1
 description Predefined level-1 role
#
role name level-2
 description Predefined level-2 role
#
role name level-3
 description Predefined level-3 role
#
role name level-4
 description Predefined level-4 role
#
role name level-5
 description Predefined level-5 role
#
role name level-6
 description Predefined level-6 role
#
role name level-7
 description Predefined level-7 role
#
role name level-8
 description Predefined level-8 role
#
role name level-9
 description Predefined level-9 role
#
role name level-10
 description Predefined level-10 role
#
role name level-11
```

```
 description Predefined level-11 role
#
role name level-12
 description Predefined level-12 role
#
role name level-13
 description Predefined level-13 role
#
role name level-14
 description Predefined level-14 role
#
user-group system
#
nat address-group 0
 address 202.20.1.6 202.20.1.6
#
nat address-group 1
 address 202.20.1.2 202.20.1.5
#
return
```

② 根据上面的信息，将 IP 地址填入表 11-2。

表 11-2 IP 地址表

设备名称	接口名称	对端设备	IP 地址	子网掩码	网关
R1	G0/0	R2	172.16.10.1	255.255.255.0	
	G0/1	S1	172.16.1.254	255.255.255.0	
	G0/2	S2	172.16.2.254	255.255.255.0	
R2	G0/1	R1	172.16.10.2	255.255.255.0	
	G0/0	Internet 服务器	202.20.1.1	255.255.255.0	
PC1			172.16.1.1	255.255.255.0	172.16.1.254
PC2			172.16.1.2	255.255.255.0	172.16.1.254
PC3			172.16.2.1	255.255.255.0	172.16.2.254
PC4			172.16.2.2	255.255.255.0	172.16.2.254
WWW 服务器			172.16.0.1	255.255.255.0	172.16.0.254
Internet 服务器			202.20.1.253	255.255.255.0	

③ 使用 disp ip routing-table 命令查看路由器 R1、R2 的路由表：

```
[R1]disp ip routing-table
Destinations : 25      Routes : 25
Destination/Mask       Proto   Pre  Cost    NextHop         Interface
0.0.0.0/32             Direct  0    0       127.0.0.1       InLoop0
127.0.0.0/8            Direct  0    0       127.0.0.1       InLoop0
127.0.0.0/32           Direct  0    0       127.0.0.1       InLoop0
127.0.0.1/32           Direct  0    0       127.0.0.1       InLoop0
127.255.255.255/32     Direct  0    0       127.0.0.1       InLoop0
```

```
172.16.0.0/24           Direct  0   0       172.16.0.254    GE5/0
172.16.0.0/32           Direct  0   0       172.16.0.254    GE5/0
172.16.0.254/32         Direct  0   0       127.0.0.1       InLoop0
172.16.0.255/32         Direct  0   0       172.16.0.254    GE5/0
172.16.1.0/24           Direct  0   0       172.16.1.254    GE0/1
172.16.1.0/32           Direct  0   0       172.16.1.254    GE0/1
172.16.1.254/32         Direct  0   0       127.0.0.1       InLoop0
172.16.1.255/32         Direct  0   0       172.16.1.254    GE0/1
172.16.2.0/24           Direct  0   0       172.16.2.254    GE0/2
172.16.2.0/32           Direct  0   0       172.16.2.254    GE0/2
172.16.2.254/32         Direct  0   0       127.0.0.1       InLoop0
172.16.2.255/32         Direct  0   0       172.16.2.254    GE0/2
172.16.10.0/24          Direct  0   0       172.16.10.1     GE0/0
172.16.10.0/32          Direct  0   0       172.16.10.1     GE0/0
172.16.10.1/32          Direct  0   0       127.0.0.1       InLoop0
172.16.10.255/32        Direct  0   0       172.16.10.1     GE0/0
202.20.1.0/24           Static  60  0       172.16.10.2     GE0/0
224.0.0.0/4             Direct  0   0       0.0.0.0         NULL0
224.0.0.0/24            Direct  0   0       0.0.0.0         NULL0
255.255.255.255/32      Direct  0   0       127.0.0.1       InLoop0
[R2]disp ip routing-table
Destinations : 17       Routes : 17
Destination/Mask        Proto   Pre Cost    NextHop         Interface
0.0.0.0/32              Direct  0   0       127.0.0.1       InLoop0
127.0.0.0/8             Direct  0   0       127.0.0.1       InLoop0
127.0.0.0/32            Direct  0   0       127.0.0.1       InLoop0
127.0.0.1/32            Direct  0   0       127.0.0.1       InLoop0
127.255.255.255/32      Direct  0   0       127.0.0.1       InLoop0
172.16.0.0/16           Static  60  0       172.16.10.1     GE0/1
172.16.10.0/24          Direct  0   0       172.16.10.2     GE0/1
172.16.10.0/32          Direct  0   0       172.16.10.2     GE0/1
172.16.10.2/32          Direct  0   0       127.0.0.1       InLoop0
172.16.10.255/32        Direct  0   0       172.16.10.2     GE0/1
202.20.1.0/24           Direct  0   0       202.20.1.1      GE0/0
202.20.1.0/32           Direct  0   0       202.20.1.1      GE0/0
202.20.1.1/32           Direct  0   0       127.0.0.1       InLoop0
202.20.1.255/32         Direct  0   0       202.20.1.1      GE0/0
224.0.0.0/4             Direct  0   0       0.0.0.0         NULL0
224.0.0.0/24            Direct  0   0       0.0.0.0         NULL0
255.255.255.255/32      Direct  0   0       127.0.0.1       InLoop0
```

④ 使用 disp nat all 命令显示 NAT 的所有信息，包括转化条目总数、NAT 配置参数、地址池中的地址及已分配的地址数：

```
[R2]disp nat all
NAT address group information:
  Totally 2 NAT address groups.
```

```
  Address group 0:
    Port range: 1-65535
    Address information:
      Start address          End address
      202.20.1.6             202.20.1.6

  Address group 1:
    Port range: 1-65535
    Address information:
      Start address          End address
      202.20.1.2             202.20.1.5

NAT internal server information:
  Totally 1 internal servers.
  Interface: GigabitEthernet0/0
    Protocol: 6(TCP)
    Global IP/port: 202.20.1.1/80
    Local IP/port : 172.16.0.1/80
    Config status : Active

NAT logging:
  Log enable              : Disabled
  Flow-begin              : Disabled
  Flow-end                : Disabled
  Flow-active             : Disabled
  Port-block-assign       : Disabled
  Port-block-withdraw     : Disabled
  Port-alloc-fail         : Disabled
  Port-block-alloc-fail   : Disabled
  Port-usage              : Disabled
  Port-block-usage        : Enabled(90%)

  Mapping mode : Address and Port-Dependent
  ACL          : ---
  Config status: Active

NAT ALG:
  DNS             : Enabled
  FTP             : Enabled
  H323            : Disabled
  ICMP-ERROR      : Enabled
  ILS             : Disabled
  MGCP            : Disabled
  NBT             : Disabled
  PPTP            : Disabled
```

```
    RTSP              : Enabled
    RSH               : Disabled
    SCCP              : Disabled
    SIP               : Disabled
    SQLNET            : Disabled
    TFTP              : Disabled
    XDMCP             : Disabled
```

根据上面的配置信息，发现路由器 R1 和 R2 上的路由及 IP 地址配置都没有问题。

2）查看 NAT 配置并根据相关知识点确定故障点的位置

① 查看 ACL 的配置：

```
[R2]disp cu
……
#
acl basic 2000
 rule 0 permit source 172.16.1.0 0.0.0.255
#
acl basic 2001
 rule 0 permit source 172.16.2.0 0.0.0.255
#
```

由于 acl basic 2000 只允许 172.16.1.0/24（PC1 使用的网段）网段的数据包通过，所以可以得出结论 acl basic 2000 配置正确，并且没有增加或遗漏需要进行地址转换的网络。

同理，acl basic 2001 只允许 172.16.2.0/24（PC3 使用的网段）网段的数据包通过，所以可以得出结论 acl basic 2001 配置正确，并且没有增加或遗漏需要进行地址转换的网络。

② 查看是否缺少静态路由配置：

```
R1:
#
 ip route-static 202.20.1.0 24 GigabitEthernet0/0 172.16.10.2
#
R2:
#
 ip route-static 172.16.0.0 16 172.16.10.1
#
```

经过查看，路由器 R1 和 R2 均配置正确。

③ 查看 NAT 地址池是否和静态地址存在重叠：

```
interface GigabitEthernet0/0
 port link-mode route
 combo enable copper
 ip address 202.20.1.1 255.255.255.0
 nat server protocol tcp global 202.20.1.1 80 inside 172.16.0.1 80
#
#
nat address-group 0
```

```
address 202.20.1.6 202.20.1.6
#
nat address-group 1
 address 202.20.1.2 202.20.1.5
#
```

根据上面的信息可以看出，202.20.1.1～202.20.1.6 这 6 个 IP 地址分别按要求归属，也没有问题。但是，地址映射缺少了两条规则，一条是领导管理人员办公网访问外网的地址转换，另一条是员工办公网访问外网的地址转换。

3）修改配置，保存配置信息，并进行测试

① 在路由器 R2 的 NAT 映射中增加新的配置：

```
nat outbound 2001 address-group 1 no-pat
nat outbound 2000 address-group 0 no-pat
```

在完成配置后，使用 disp cu 命令进行查看。

在终端中 ping 外网服务器（Internet），结果显示成功。

查看路由器 R2 的 NAT 地址映射表：

```
[R2]disp nat session bri
Slot 0:
Protocol        Source IP/port          Destination IP/port      Global IP/port
ICMP            172.16.2.1/162          202.20.1.253/2048        202.20.1.2/0
Total sessions found: 1
```

根据上面的信息可知，地址映射成功。保存修改后的配置。

② 在完成配置后，重新对终端进行互联互通测试，结果如表 11-3 所示。

表 11-3 终端互联互通测试结果

测试序号	本端设备	对端设备	测试方法	预期测试结果	实际测试结果	是否发生故障
1	PC1	Server 服务器	ping 命令	成功	成功	否
2	PC2	Server 服务器	ping 命令	成功	成功	否
3	PC3	Server 服务器	ping 命令	成功	成功	否
4	PC4	Server 服务器	ping 命令	成功	成功	否

使用 ping 命令，分别测试 PC1、PC2、PC3、PC4 与 Internet 服务器的连通性。

在 PC1 中 ping 公网地址：

```
<PC1>ping 202.20.1.1
Ping 202.20.1.1 (202.20.1.1): 56 data bytes, press CTRL_C to break
56 bytes from 202.20.1.1: icmp_seq=0 ttl=253 time=3.000 ms
56 bytes from 202.20.1.1: icmp_seq=1 ttl=253 time=2.000 ms
56 bytes from 202.20.1.1: icmp_seq=2 ttl=253 time=2.000 ms
56 bytes from 202.20.1.1: icmp_seq=3 ttl=253 time=2.000 ms
```

在 PC2 中 ping 公网地址：

```
<PC2>ping 202.20.1.1
Ping 202.20.1.1 (202.20.1.1): 56 data bytes, press CTRL_C to break
56 bytes from 202.20.1.1: icmp_seq=0 ttl=253 time=3.000 ms
56 bytes from 202.20.1.1: icmp_seq=1 ttl=253 time=2.000 ms
```

```
56 bytes from 202.20.1.1: icmp_seq=2 ttl=253 time=2.000 ms
56 bytes from 202.20.1.1: icmp_seq=3 ttl=253 time=2.000 ms
56 bytes from 202.20.1.1: icmp_seq=4 ttl=253 time=2.000 ms
```

在 PC3 中 ping 公网地址：

```
<PC3>ping 202.20.1.1
Ping 202.20.1.1 (202.20.1.1): 56 data bytes, press CTRL_C to break
56 bytes from 202.20.1.1: icmp_seq=0 ttl=253 time=3.000 ms
56 bytes from 202.20.1.1: icmp_seq=1 ttl=253 time=2.000 ms
56 bytes from 202.20.1.1: icmp_seq=2 ttl=253 time=2.000 ms
56 bytes from 202.20.1.1: icmp_seq=3 ttl=253 time=2.000 ms
56 bytes from 202.20.1.1: icmp_seq=4 ttl=253 time=2.000 ms
```

在 PC4 中 ping 公网地址：

```
<PC4>ping 202.20.1.1
Ping 202.20.1.1 (202.20.1.1): 56 data bytes, press CTRL_C to break
56 bytes from 202.20.1.1: icmp_seq=0 ttl=253 time=3.000 ms
56 bytes from 202.20.1.1: icmp_seq=1 ttl=253 time=2.000 ms
56 bytes from 202.20.1.1: icmp_seq=2 ttl=253 time=2.000 ms
56 bytes from 202.20.1.1: icmp_seq=3 ttl=253 time=2.000 ms
56 bytes from 202.20.1.1: icmp_seq=4 ttl=253 time=2.000 ms Router#
```

由表 11-3 可知，有 4 项测试成功，由此确认故障现象和故障分析完全吻合。在经过故障点分析、故障点确定，以及分步骤修改配置后，成功完成故障排除工作。

4）整理新的配置文档

在故障排除后，保存所有路由器的配置信息，并更新书面的记录材料，确保书面文档和实际配置保持一致，以确保下次配置正常使用。

11.3 相关知识准备

知识准备

为了能够深入地分析故障点，读者应了解 NAT 的相关知识。

1）什么是 NAT

当前的 Internet 主要基于 IPv4 协议，用户访问 Internet 的前提条件是拥有属于自己的 IPv4 地址。IPv4 地址共 32 位，理论上支持约 40 亿的地址空间，但随着 Internet 用户的快速增长，加上地址分配不均等因素，很多国家已经陷入 IP 地址不足的困境。

为了解决 IPv4 地址不足的问题，IETF（Internet Engineering Task Force，因特网工程任务组）提出了 NAT（Network Address Translation，网络地址转换）解决方案。IP 地址分为公有地址和私有地址。公有地址由 IANA 统一分配，用于 Internet 通信；私有地址可以自由分配，用于私有网络内部通信。NAT 技术的主要作用是将 IP 数据包报头中的 IP 地址转换为另一个 IP 地址，即将私有地址转换为公有地址，使私有网络中的主机可以通过共享少量公有地址访问 Internet。NAT 技术在一定程度上解决了可用公有 IP 地址不足的问题。

然而，NAT 只是一种过渡技术。从根本上解决地址供需问题的方法是采用支持更大地址空间的下一代 IP 技术，即 IPv6 协议。IPv6 协议提供了几乎取之不尽的地址空间，是下一代 Internet 的协议基础。

2）与 NAT 相关的常用术语

① 公网：使用 IANA 分配的公有 IP 地址空间的网络，或者在互联的两个网络中不需要做地址转换的一方。在讨论 NAT 时，公网常常被称为全局网络（Global Network）或外网（External Network）。相应地，公网节点使用的地址被称为公有 IP 地址（Public IP Address）或全局地址（Global Address）。

② 私网：使用独立于外部网络的私有 IP 地址空间的内部网络，或者在互联的两个网络中，需要做地址转换的一方。在讨论 NAT 时，私网常常被称为本地网络（Local Network）或内网（Internal Network）。相应地，私网节点使用的地址被称为私有 IP 地址（Private IP Address）或本地地址（Local Address）。

③ NAT 设备（NAT Device）：介于公网与私网之间的设备，负责执行公有地址与私有地址的转换。通常由一台路由器来完成这个任务。

④ TU Port：与某个 IP 地址相关联的 TCP/UDP 接口，如 HTTP 的 TU Port 为 80。

⑤ 地址池（Address Pool）：一般是公有地址的集合。在配置动态地址转换后，NAT 设备会从地址池中动态为私网用户分配公有地址。

3）NAT 转换的两种类型

① 动态 NAT：使用公有地址池，并以先到先得的原则分配这些地址。当使用私有 IP 地址的主机请求访问 Internet 时，动态 NAT 会从地址池中选择一个未被其他主机使用的 IP 地址进行分配，这就是前面介绍的地址映射。

② 静态 NAT：使用本地地址与全局地址的一对一映射，这些映射保持不变。对必须使用固定地址以便能够从 Internet 访问的 WWW 服务器或主机来说，静态 NAT 很有用。这些内部主机可能是企业服务器或网络设备。

无论是动态 NAT 还是静态 NAT，都必须有足够的 IP 地址，以便为同时发生的每个用户会话分配一个 IP 地址。

11.4 项目小结

本项目主要针对 NAT 配置进行故障排除，以理论化故障排除思路为指导，以任务式驱动为方法，形象地再现了 NAT 的原理和操作过程，为未来实际环境的 NAT 故障排除提供了良好的方法和操作步骤。

素质拓展：换道超车 把根留住

考虑网络应用网页访问的工作原理，分析 DNS 的工作过程。无论是互联网运营商还是普通网民，都非常重视和关心域名根服务器的引入问题。

2017年11月26日,中共中央办公厅、国务院办公厅印发《推进互联网协议第六版(IPv6)规模部署行动计划》,开始IPv6的全国战略部署。

第52次《中国互联网络发展状况统计报告》显示,"在网络基础资源方面,截至2023年6月,我国域名总数为3024万个;IPv6地址数量为68 055块/32,IPv6活跃用户数达7.67亿;互联网宽带接入端口数量达11.1亿个;光缆线路总长度达6196万公里。在移动网络发展方面,截至6月,我国移动电话基站总数达1129万个,其中累计建成开通5G基站293.7万个,占移动基站总数的26%;移动互联网累计流量达1423亿GB,同比增长14.6%;移动互联网应用蓬勃发展,国内市场上监测到的活跃APP数量达260万款,进一步覆盖网民日常学习、工作、生活。在物联网发展方面,截至6月,三家基础电信企业发展蜂窝物联网终端用户21.23亿户,较2022年12月净增2.79亿户,占移动网终端连接数的比重为55.4%,万物互联基础不断夯实。"

我们时常会看到一些关于驻外使馆官网域名变更的新闻,如中国驻瑞士大使馆网站将域名从china-embassy.org更改为china-embassy.gov.cn。在一个域名后面添加"cn"后缀,表明该国家的DNS解析服务已经实现了全球化部署。

增值服务

随着访问量的增加,当一台服务器无法满足需求时,就需要采用负载均衡技术,将大量的访问合理地分配到多台服务器上。

例如,现有这样一个网络环境:局域网以2Mbps DDN专线接入Internet,路由器选用安装了广域网模块的H3C MSR 2600;内部网络使用的IP地址段为10.1.1.1~10.1.3.254,局域网端口Ethernet 0/1的IP地址为10.1.1.1/22;网络分配的合法IP地址范围为202.110.198.80~202.110.198.87,连接ISP的端口Serial 0/0的IP地址为202.110.198.81,子网掩码为255.255.255.248。要求网络内部的所有计算机都能够访问Internet,并且在3台Web服务器和2台FTP服务器中实现负载均衡。

案例分析:既然要求网络内的所有计算机都能够接入Internet,而合法IP地址又只有5个可用,那么可以采用端口复用地址转换方式。但是,由于服务器的访问量过大(或者服务器的性能较差),不得不使用多台服务器进行负载均衡。因此,必须将一个合法IP地址转换成多个内部IP地址,并采用轮询方式来分担每台服务器的访问压力。

定义局域端口:
```
interface gigabitethernet 0/0/1
ip adderss 10.1.1.1 255.255.252.0  //定义局域网端口的IP地址
duplex auto
speed auto
nat inside
```

定义广域端口:
```
interface serial 0/0
ip address 202.110.198.81 255.255.255.248  //定义广域网端口的IP地址
duplex auto
```

```
speed auto
nat outside
```

配置服务器列表：
```
system-view
ip pool server-list1    //2 台 FTP 服务器
 server 10.1.1.8
 server 10.1.1.9
ip pool server-list2    //3 台 Web 服务器
 server 10.1.1.2
 server 10.1.1.3
 server 10.1.1.4
```

创建服务器组：
```
server-group name my-server-group1
 server-address ip pool server-list1
server-group name my-server-group2
 server-address ip pool server-list2
```

创建访问策略：
```
acl number 2000
 rule 10 permit source any
```

设置服务器组和轮询方式：
```
webauth policy global
 lb group my-server-group1
 lb method round-robin
 lb group my-server-group2
 lb method round-robin
```

应用访问策略：
```
interface Serial 0/0
 traffic-policy 2000 outbound
```

11.5 课后实训

项目内容：某公司的分公司 A 要访问总公司的网站，分公司 A 内网使用 192.168.1.0/24 网段。总公司向当地 ISP 申请了 3 个公有 IP 地址，分别是 222.1.0.1/24、222.1.0.2/24、222.1.0.3/24。其中，222.1.0.1/24 作为总路由器出口地址，222.1.0.2/24 作为网站服务器地址。

目前分公司 A 路由器的接口配置及 NAT 基本配置已经完成，但其他子公司都可以正常访问总公司的网站，唯独分公司 A 无法访问总公司的网站。请按如下操作步骤排除故障。

① 根据要求检查故障现象。
② 根据故障现象收集故障信息。
③ 利用结构化故障排除方法完成故障定位。
④ 修改故障配置并说明故障原因。
⑤ 更新配置文档。

项目 12

Telnet 协议应用的故障排除

内容介绍

某公司的总部在上海,旗下有两个办事处,并且办事处分布在不同的城市。该公司想要在总部通过 4 台 PC 远程管理旗下各个办事处的网络,实现远程集中管理。为实现这一目标,需要使用 Telnet 协议。

任务安排

任务 1　针对业务需求实现 Telnet 远程登录网络配置
任务 2　进行网络配置更新过程中的故障分析与排除

学习目标

◇ 了解 Telnet 远程登录应用常见故障的产生原因
◇ 掌握故障排除的思路
◇ 学会结构化故障排除方法
◇ 学会 Telnet 远程登录应用相关故障排除及文档更新的方法

素质目标

增强网络安全意识,承担维护网络安全的责任,自觉遵守法规与伦理,提高自身社会责任感。

12.1 Telnet 协议配置分析与实施

发现故障

网络管理员李工在接到公司网络 Telnet 远程登录更新任务后，根据公司要求，制定了网络实施方案，主要操作如下。

① 使各个办事处的路由器均开启 Telnet 协议。
② 配置 Telnet 远程登录管理地址。
③ 在终端中测试 Telnet 远程登录。

公司网络拓扑结构如图 12-1 所示。

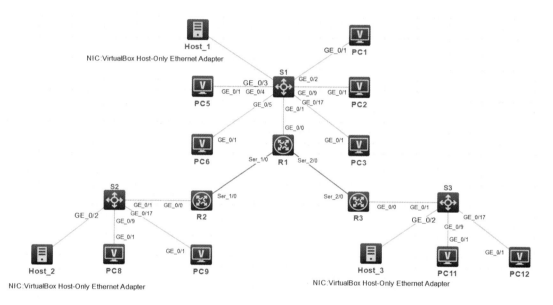

图 12-1　公司网络拓扑结构

1) 使各个办事处的路由器均开启 Telnet 协议

对路由器 R1 进行配置：

```
//开启 Telnet 协议，并配置用户名与密码
[R1]local-user test
New local user added.
[R1-luser-manage-test]password simple 123
[R1-luser-manage-test]service-type telnet
[R1-luser-manage-test]authorization-attribute user-role level-0
[R1]telnet server enable
[R1]line vty 0 4
[R1-line-vty0-4]authentication-mode scheme
```

在路由器 R2、R3 上执行与路由器 R1 相同的操作，这里不再赘述。

2) 配置 Telnet 远程登录管理地址

配置路由器 R1 的 Telnet 远程登录管理地址：

```
interface Loopback0
```

```
ip address 1.1.1.1 255.255.255.255
```
配置路由器 R2 的 Telnet 远程登录管理地址：
```
interface Loopback0
ip address 2.2.2.2 255.255.255.255
```
配置路由器 R3 的 Telnet 远程登录管理地址：
```
interface Loopback0
ip address 3.3.3.3 255.255.255.255
```

3）在终端中测试 Telnet 远程登录

配置完成后，在远程主机 Host_1、Host_2 和 Host_3 中进行测试，发现如下问题：一是，Host_1、Host_2 和 Host_3 可以远程登录路由器 R3，但是无法进入特权模式，并且提示"没有设置密码"，而从上面的配置看，已经设置了 Telnet 远程登录密码；二是，当 3 台 PC 同时 Telnet 远程登录路由器 R2 或 R3 时，只能同时登录 2 个用户，总有 2 个 Telnet 用户无法登录。接下来对以上问题进行分析与排除。

12.2 Telnet 协议配置故障分析与排除

排除故障

1. 故障分析方法

根据结构化故障排除思路，严格执行故障排除的操作步骤。首先，确定故障现象并进行详细记录。其次，收集设备信息。本项目主要收集交换机的配置信息、IP 地址、接口信息、密码信息等。再次，在收集信息后，结合 Telnet 的实现原理进行综合分析，确定并罗列可能存在的故障点。最后，针对故障点分析出最有可能的故障原因，并对这个原因进行故障排除。本次故障排除可以分解为以下几个任务。

① 查看所有路由器的配置。
② 查看路由器的 Telnet 配置。
③ 根据相关知识点确定故障点的位置。

2. 分析故障点

从 Telnet 远程登录的实现及技术原理进行分析，可能存在以下故障点。
① 管理地址配置有错误。
② 密码设置有错误。
③ 只启用了 Telnet 协议，却没有设置远程登录认证方式。
④ 在 Telnet 远程登录设置中，设置的 Telnet 用户数小于实际 Telnet 用户数。

3. 确认故障所需设备

① 一台装有超级终端软件或 Telnet 软件的计算机，同时确定访问所需的用户名和口令。
② 配置线缆。

③ 笔和纸，用于记录相关信息。

4．实施步骤

1）查看路由器的管理地址

查看路由器 R1 的管理地址：

```
[R1]disp ip inter bri
*down: administratively down
(s): spoofing (l): loopback
Interface            Physical  Protocol  IP Address       Description
GE0/0                up        up        192.168.1.1      --
GE0/1                down      down      --               --
GE0/2                down      down      --               --
GE5/0                down      down      --               --
GE5/1                down      down      --               --
GE6/0                down      down      --               --
GE6/1                down      down      --               --
Loop0                up        up(s)     1.1.1.1          --
Ser1/0               up        up        192.168.3.1      --
Ser2/0               down      down      192.168.5.1      --
Ser3/0               down      down      --               --
Ser4/0               down      down      --               --
```

查看路由器 R2 的管理地址：

```
[R2]disp ip inter bri
*down: administratively down
(s): spoofing (l): loopback
Interface            Physical  Protocol  IP Address       Description
GE0/0                up        up        192.168.2.1      --
GE0/1                down      down      --               --
GE0/2                down      down      --               --
GE5/0                down      down      --               --
GE5/1                down      down      --               --
GE6/0                down      down      --               --
GE6/1                down      down      --               --
Loop0                up        up(s)     2.2.2.2          --
Ser1/0               up        up        192.168.3.2      --
Ser2/0               down      down      --               --
Ser3/0               down      down      --               --
Ser4/0               down      down      --               --
```

查看路由器 R3 的管理地址：

```
[R3]disp ip inter bri
*down: administratively down
(s): spoofing (l): loopback
Interface            Physical  Protocol  IP Address       Description
GE0/0                up        up        192.168.4.1      --
```

GE0/1	down	down	--	--
GE0/2	down	down	--	--
GE5/0	down	down	--	--
GE5/1	down	down	--	--
GE6/0	down	down	--	--
GE6/1	down	down	--	--
Loop0	up	up(s)	3.3.3.3	--
Ser1/0	down	down	--	--
Ser2/0	up	up	192.168.5.2	--
Ser3/0	down	down	--	--

经过检查，路由器的管理地址配置没有错误。

2）重设密码并排除密码设置错误

因为密码在系统中加密显示，所以可以通过重设密码来解决可能的错误。

重设路由器 R1 的密码：

```
[R1-luser-manage-test]password simple 123
```

重设路由器 R2、R3 的密码：

```
[R2-luser-manage-test]password simple 123
[R3-luser-manage-test]password simple 123
```

3）查看是否只启用了 Telnet 协议，却没有设置远程登录认证方式

查看路由器 R2 的所有配置：

```
[R2] disp cu
#
 version 7.1.075, Alpha 7571
#
 sysname R2
#
 telnet server enable
#
rip 1
 network 2.0.0.0
 network 192.168.2.0
 network 192.168.3.0
#
 system-working-mode standard
 xbar load-single
 password-recovery enable
 lpu-type f-series
#
vlan 1
#
interface Serial1/0
 ip address 192.168.3.2 255.255.255.0
#
interface Serial2/0
```

```
#
interface Serial3/0
#
interface Serial4/0
#
interface NULL0
#
interface LoopBack0
 ip address 3.3.3.3 255.255.255.0
#
interface GigabitEthernet0/0
 port link-mode route
 combo enable copper
 ip address 192.168.2.1 255.255.255.0
#
interface GigabitEthernet0/1
 port link-mode route
 combo enable copper
#
interface GigabitEthernet0/2
 port link-mode route
 combo enable copper
#
interface GigabitEthernet5/0
 port link-mode route
 combo enable copper
#
interface GigabitEthernet5/1
 port link-mode route
 combo enable copper
#
interface GigabitEthernet6/0
 port link-mode route
 combo enable copper
#
interface GigabitEthernet6/1
 port link-mode route
 combo enable copper
#
 scheduler logfile size 16
#
line class aux
 user-role network-operator
#
line class console
```

```
 user-role network-admin
#
line class tty
 user-role network-operator
#
line class vty
 user-role network-operator
#
line aux 0
 user-role network-operator
#
line con 0
 user-role network-admin
#
line vty 0 1
 authentication-mode scheme
 user-role network-operator
#
line vty 2 63
 user-role network-operator
#
domain name system
#
domain default enable system
#
role name level-0
 description Predefined level-0 role
#
role name level-1
 description Predefined level-1 role
#
role name level-2
 description Predefined level-2 role
#
role name level-3
 description Predefined level-3 role
#
role name level-4
 description Predefined level-4 role
#
role name level-5
 description Predefined level-5 role
#
role name level-6
 description Predefined level-6 role
```

```
#
role name level-7
 description Predefined level-7 role
#
role name level-8
 description Predefined level-8 role
#
role name level-9
 description Predefined level-9 role
#
role name level-10
 description Predefined level-10 role
#
role name level-11
 description Predefined level-11 role
#
role name level-12
 description Predefined level-12 role
#
role name level-13
 description Predefined level-13 role
#
role name level-14
 description Predefined level-14 role
#
user-group system
#
local-user test class manage
 password hash $h$6$Eh5I3DJLg6t9qIFe$rNr8EljVNvHKzWQKqnl0N0HEPTFHxCVYuD
A8tQ2j2LFwZOb1RO1Dq/JNJM7jxtiDvAICLb3uogJ2Wb5TaBtTFQ==
 service-type telnet
 authorization-attribute user-role level-0
 authorization-attribute user-role network-operator
#
return
```

查看路由器 R3 的所有配置：

```
[R3] disp cu
#
 version 7.1.075, Alpha 7571
#
 sysname R3
#
 telnet server enable
#
rip 1
```

```
 network 3.0.0.0
 network 192.168.4.0
 network 192.168.5.0
#
 system-working-mode standard
 xbar load-single
 password-recovery enable
 lpu-type f-series
#
vlan 1
#
interface Serial1/0
#
interface Serial2/0
 ip address 192.168.5.2 255.255.255.0
#
interface Serial3/0
#
interface Serial4/0
#
interface NULL0
#
interface LoopBack0
 ip address 3.3.3.3 255.255.255.0
#
interface GigabitEthernet0/0
 port link-mode route
 combo enable copper
 ip address 192.168.4.1 255.255.255.0
#
interface GigabitEthernet0/1
 port link-mode route
 combo enable copper
#
interface GigabitEthernet0/2
 port link-mode route
 combo enable copper
#
interface GigabitEthernet5/0
 port link-mode route
 combo enable copper
#
interface GigabitEthernet5/1
 port link-mode route
 combo enable copper
```

```
#
interface GigabitEthernet6/0
 port link-mode route
 combo enable copper
#
interface GigabitEthernet6/1
 port link-mode route
 combo enable copper
#
 scheduler logfile size 16
#
line class aux
 user-role network-operator
#
line class console
 user-role network-admin
#
line class tty
 user-role network-operator
#
line class vty
 user-role network-operator
#
line aux 0
 user-role network-operator
#
line con 0
 user-role network-admin
#
line vty 0 1
 authentication-mode scheme
 user-role network-operator
#
line vty 2 63
 user-role network-operator
#
domain name system
#
 domain default enable system
#
role name level-0
 description Predefined level-0 role
#
role name level-1
 description Predefined level-1 role
```

```
#
role name level-2
 description Predefined level-2 role
#
role name level-3
 description Predefined level-3 role
#
role name level-4
 description Predefined level-4 role
#
role name level-5
 description Predefined level-5 role
#
role name level-6
 description Predefined level-6 role
#
role name level-7
 description Predefined level-7 role
#
role name level-8
 description Predefined level-8 role
#
role name level-9
 description Predefined level-9 role
#
role name level-10
 description Predefined level-10 role
#
role name level-11
 description Predefined level-11 role
#
role name level-12
 description Predefined level-12 role
#
role name level-13
 description Predefined level-13 role
#
role name level-14
 description Predefined level-14 role
#
user-group system
#
local-user R3 class manage
 password hash $h$6$Eh5I3DJLg6t9qIFe$rNr8EljVNvHKzWQKqnl0N0HEPTFHxCVYuD
A8tQ2j2LFwZOb1ROlDq/JNJM7jxtiDvAICLb3uogJ2Wb5TaBtTFQ==
```

```
 service-type telnet
 authorization-attribute user-role level-0
 authorization-attribute user-role network-operator
#
return
```

经过检查，发现路由器 R3 的账号设置错误，所以 Host_1、Host_2、Host_3 在进行远程登录时不能进入特权模式。

4）查看在 Telnet 远程登录设置中，设置的 Telnet 用户数是否小于实际 Telnet 用户数

从步骤 3）可以发现，路由器 R2、R3 的 Telnet 远程登录设置如下：

```
line vty 0 1
 authentication-mode scheme
 user-role network-operator
#
```

经过检查，发现路由器 R2、R3 上的 Telnet 远程登录用户数均为 2，而总部要求同时允许 4 个 Telnet 用户登录，所以用户数设置限制了可以同时登录的用户数量，导致其中两个 Telnet 用户无法实现远程登录。接下来修改配置，把用户数设置为 5 个，并重新进行测试，结果显示每个用户都能成功登录。

5）修改配置，并保存配置信息

① 在路由器 R3 中创建一个用户，用户名为 test：

```
[R3]local-user test
```

为该用户创建登录时的认证密码，密码为 123，这里使用 password 命令来指定密码配置方式（密码配置方式为 simple，表示以明文方式配置密码）：

```
[R3-luser-manage-test]password simple 123
```

设置该用户使用 Telnet 服务类型，以及设置该用户的用户角色 user-role 为 level-0（level-number 中的 number 对应用户角色的级别，数值越小，用户的权限级别越低）：

```
[R3-luser-manage-test]service-type telnet
[R3-luser-manage-test]authorization-attribute user-role level-0
[R3-luser-manage-test]quit
```

② 打开 Telnet 服务：

```
[R3]telnet server enable
```

③ 配置对 Telnet 用户使用默认的本地认证。

远程登录用户数的配置：

```
[R3]line vty 0 4
```

交换机可以采用本地或第三方服务器来对用户进行认证，这里使用本地认证授权方式（认证模式为 scheme）：

```
[R3-line-vty0-4]authentication-mode scheme
```

由 Host_1 连接路由器 R3（管理地址为 3.3.3.3），登录成功，如图 12-2 所示。

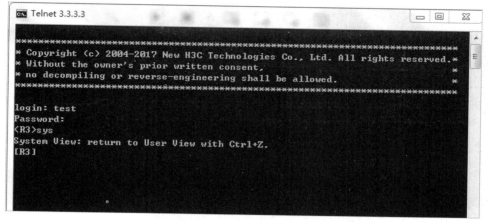

图 12-2　远程登录界面

从图 12-2 中可知，测试成功，由此确认故障现象和故障分析完全吻合。在经过故障点分析、故障点确定，以及分步骤修改配置后，成功完成故障排除工作。

6）整理新的配置文档

在故障排除后，保存所有路由器的配置信息，并更新书面的记录材料，确保书面文档和实际配置保持一致，以确保下次配置正常使用。

12.3　相关知识准备

知识准备

为了能够深入地分析故障点，读者应了解 Telnet 的相关知识。

Telnet 是一个远程终端协议。Telnet 应用虽然方便了用户进行远程登录，但是也给很多非法用户提供了一种入侵手段和后门。所以，在开启 Telnet 应用，实现便捷维护的同时，我们要了解 Telnet，关上非法使用 Telnet 的"大门"。

远程登录是指用户可以坐在联网的主机前，登录远距离的另一台联网主机，成为这台主机的终端。远程登录使用户可以方便地操纵另一端的主机，就像它在身边一样。通过远程登录，本地主机可以与网络上的另一台远程主机建立连接，并进行程序交互。进行远程登录的用户被称为本地用户，本地用户登录的系统被称为远地系统。主机分时系统允许多个用户同时登录使用一台设备。为了保证系统的安全和方便记账，系统要求每个用户拥有一个单独的账号作为登录标识，并为每个用户分配一个口令。用户在使用该系统之前需要输入其标识和口令。

使用 Telnet 协议进行远程登录时需要满足以下条件：本地主机必须安装包含 Telnet 协议的客户程序；必须知道远程主机的 IP 地址；必须知道登录账号和口令。

Telnet 远程登录服务分为以下 4 个过程。

① 本地主机与远程主机建立连接。该过程实际上是建立一个 TCP 连接。用户必须知道远程主机的 IP 地址或域名。

② 将本地主机上输入的用户名和口令及以后输入的任何命令或字符以 NVT（Network

Virtual Terminal，网络虚拟终端）格式传送到远程主机上。

③ 将远程主机输出的 NVT 格式的数据转化为本地接受的格式，并发送给本地主机。这些数据包括输入命令的回显和命令执行结果。

④ 本地主机断开与远程主机的连接。

Telnet 协议是 TCP/IP 协议簇中的一员，也是 Internet 远程登录服务的标准协议和主要方式。Telnet 协议为用户提供了一种在本地主机上完成远程主机工作的方式。用户可以通过 Telnet 程序在本地主机中输入命令，就像直接在远程主机的控制台上输入一样，这些命令会在远程主机上运行。

虽然 Telnet 服务较为简单、使用也很方便，但在注重安全的现代网络技术中，Telnet 服务并不被推荐使用。原因在于 Telnet 是一个明文传送协议，它会将用户的所有内容（包括用户名和密码）以明文形式在互联网上传送，存在一定的安全隐患。因此，许多服务器都会选择禁用 Telnet 服务。如果用户要使用 Telnet 服务进行远程登录，在使用前应在远程主机上检查并设置允许 Telnet 服务的功能，添加远程登录用户和口令，并且采用 SSH 协议为远程登录会话等网络服务提供安全保障。

12.4　项目小结

本项目主要针对 Telnet 协议配置进行故障排除，以理论化故障排除思路为指导，以任务式驱动为方法，形象地再现了 Telnet 协议的工作条件，为未来实际环境的 Telnet 协议故障排除提供了良好的方法和操作步骤。

素质拓展：数字中国　利国利民

习近平总书记高度重视互联网与数字经济发展，做出了一系列重要论述。以下罗列部分论述。

当今时代，以信息技术为核心的新一轮科技革命正在孕育兴起，互联网日益成为创新驱动发展的先导力量，深刻改变着人们的生产生活，有力推动着社会发展。

——2014 年 11 月 19 日，致首届互联网大会的贺词

要运用大数据促进保障和改善民生。大数据在保障和改善民生方面大有作为。要坚持以人民为中心的发展思想，推进"互联网+教育"、"互联网+医疗"、"互联网+文化"等，让百姓少跑腿、数据多跑路，不断提升公共服务均等化、普惠化、便捷化水平。

——2017 年 12 月 8 日，在中共中央政治局第二次集体学习时的讲话

要推动互联网、大数据、人工智能和实体经济深度融合，加快制造业、农业、服务业数字化、网络化、智能化。

——2018 年 4 月 20 日，在全国网络安全和信息化工作会议上的讲话

当今世界，正在经历一场更大范围、更深层次的科技革命和产业变革。互联网、大数

据、人工智能等现代信息技术不断取得突破，数字经济蓬勃发展，各国利益更加紧密相连。为世界经济发展增添新动能，迫切需要我们加快数字经济发展，推动全球互联网治理体系向着更加公正合理的方向迈进。

——2018年11月7日，致第五届世界互联网大会的贺信

中国正在大力建设"数字中国"，在"互联网+"、人工智能等领域收获一批创新成果。分享经济、网络零售、移动支付等新技术新业态新模式不断涌现，深刻改变了中国老百姓生活。

——2018年11月18日，在亚太经合组织第二十六次领导人非正式会议上的发言

要探索"区块链+"在民生领域的运用，积极推动区块链技术在教育、就业、养老、精准脱贫、医疗健康、商品防伪、食品安全、公益、社会救助等领域的应用，为人民群众提供更加智能、更加便捷、更加优质的公共服务。

——2019年10月24日，在中共中央政治局第十八次集体学习时的讲话

运用大数据、云计算、区块链、人工智能等前沿技术推动城市管理手段、管理模式、管理理念创新，从数字化到智能化再到智慧化，让城市更聪明一些、更智慧一些，是推动城市治理体系和治理能力现代化的必由之路，前景广阔。

——2020年3月31日，在浙江杭州城市大脑运营指挥中心考察调研时的讲话

增值服务

在每次项目修复完成后，售后工程人员必须给甲方提供工程结单。经过网络工程项目实践后，我们需要认真负责、具备钻研创新精神和服务意识，编写一份完整的工程结项报告，以提供给用户一个满意的定制文档。

12.5 课后实训

项目内容：网络管理员王工所在公司要求公司网络内的所有路由器均实现远程登录。其中，路由器A连接终端使用的网段为192.168.10.0/24，出口IP地址为202.111.1.1/24。王工通过Console口登录各路由器并做好配置后，发现其他路由器均可实现远程登录，唯独路由器A无法实现远程登录。请按如下操作步骤排除故障。

① 根据要求检查故障现象。
② 根据故障现象收集故障信息。
③ 利用结构化故障排除方法完成故障定位。
④ 修改故障配置并说明故障原因。
⑤ 更新配置文档。

项目 13

网络自动化运维

内容介绍

通过项目建设，甲公司新建的办公大楼的现有网络架构已经能够满足日常办公需求，项目转入运维阶段。为满足运维需求，公司在网管计算机上预装了 CentOS 7.0 操作系统，计划使用 Python 进行网络自动运维。因此，网络管理员被分配了如下任务。

（1）在项目转入运维阶段后，公司管理员应马上修改所有网络设备的管理密码。

（2）每天凌晨 1 点对所有网络设备执行一次自动备份配置。

任务安排

任务 1　在项目转入运维阶段后，马上修改所有网络设备的管理密码

任务 2　每天凌晨 1 点对所有网络设备执行一次自动备份配置

学习目标

◇ 了解 Python 运维的常用库和常用语法

◇ 掌握通过 Python 代码管理网络设备的方法

◇ 掌握通过 Python 代码备份网络设备配置的方法

素质目标

具备工作热情、创新意识，拥有较强的岗位适应能力和可持续发展能力。

项目 13　网络自动化运维

13.1　公司实际需求分析

需求分析

甲公司实际需求如下。

（1）根据公司实际需求，网络管理员需要对这批网络设备的管理密码进行批量修改，并定期对设备配置做备份。在本项目中，可以先在网管计算机上使用 Python 脚本来加载 Paramiko 模块，再通过 SSH 协议批量修改网络设备的登录密码。

（2）公司还有定期备份配置这样的工作计划，可以调用网管计算机上的计划任务程序，让计算机按计划执行特定的 Python 脚本来实现。

13.2　本项目实施具体工作任务

根据公司需求，明确以下任务。

（1）使用 Python 完成自动修改网络设备的管理密码。

（2）使用 Python 和计划任务程序完成网络设备的每天自动备份。

13.3　项目背景

甲公司有三大办公区域，各区域之间使用路由器进行互联。技术部、财务部、市场部分别使用路由器 R1、R2、R3，并需要配置单区域 OSPF 动态路由，以确保所有计算机之间能够互相访问。为了安全，公司需要统一修改所有网络设备的管理密码，并设置每天自动备份。网络拓扑结构如图 13-1 所示，具体要求如下。

（1）在各路由器之间配置 OSPF 路由，实现网络的互联互通。

（2）在各路由器上启用 SNMP，实现对路由器进行网络管理，统一修改所有设备的管理密码并设置每天自动备份。

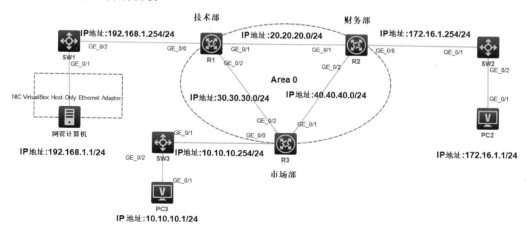

图 13-1　网络拓扑结构

13.3.1 项目规划设计

三大办公区域之间能够相互通信，为路由器配置单区域OSPF动态路由，使所有计算机均能够互相访问。为所有网络设备开启SSH，在技术部设置一台网管计算机，以确保网管计算机与网络设备之间的正常通信。

配置步骤如下。
（1）配置路由器的接口。
（2）部署单区域OSPF网络。
（3）在路由器上配置SSH登录。
（4）在网管计算机上安装模块。
（5）编写Python修改密码脚本。
（6）编写Python备份脚本。
（7）配置计划任务。
（8）配置各计算机的IP地址。

具体规划如表13-1和表13-2所示。

表13-1 IP地址规划表

设备	接口	IP地址
R1	GE_0/0	192.168.1.254/24
R1	GE_0/1	20.20.20.1/24
R1	GE_0/2	30.30.30.1/24
R2	GE_0/0	172.16.1.254/24
R2	GE_0/1	20.20.20.10/24
R2	GE_0/2	40.40.40.10/24
R3	GE_0/0	10.10.10.254/24
R3	GE_0/1	40.40.40.1/24
R3	GE_0/2	30.30.30.10/24
网管计算机	NIC:Host-Only	192.168.1.1/24
PC2	GE_0/1	172.16.1.1/24
PC3	GE_0/1	10.10.10.1/24

表13-2 端口规划表

本端设备	本端接口	对端设备	对端接口
R1	GE_0/0	SW1	GE_0/2
R1	GE_0/1	R2	GE_0/1
R1	GE_0/2	R3	GE_0/2

续表

本端设备	本端接口	对端设备	对端接口
R2	GE_0/0	SW2	GE_0/1
R2	GE_0/1	R1	GE_0/1
R2	GE_0/2	R3	GE_0/1
R3	GE_0/0	SW3	GE_0/1
R3	GE_0/1	R2	GE_0/2
R3	GE_0/2	R1	GE_0/2
SW1	GE_0/2	R1	GE_0/0
SW1	GE_0/1	网管计算机	NIC:Host-Only
SW2	GE_0/1	R2	GE_0/0
SW2	GE_0/2	PC2	GE_0/1
SW3	GE_0/1	R3	GE_0/0
SW3	GE_0/2	PC3	GE_0/1
网管计算机	NIC:Host-Only	SW1	GE_0/1
PC2	GE_0/1	SW2	GE_0/1
PC3	GE_0/1	SW3	GE_0/2

13.3.2 项目实施

1）配置路由器的接口

① 配置路由器 R1 的接口：

```
<H3C>sys
System View: return to User View with Ctrl+Z.
[H3C]sysname R1
[R1]interface GigabitEthernet 0/0
[R1-GigabitEthernet0/0]ip address 192.168.1.254 24
[R1-GigabitEthernet0/0]quit
[R1]interface GigabitEthernet 0/1
[R1-GigabitEthernet0/1]ip address 20.20.20.1 24
[R1-GigabitEthernet0/1]quit
[R1]interface GigabitEthernet 0/2
[R1-GigabitEthernet0/2]ip address 30.30.30.1 24
[R1-GigabitEthernet0/2]quit
```

② 配置路由器 R2 的接口：

```
<H3C>sys
System View: return to User View with Ctrl+Z.
[H3C]sysname R2
[R2]interface GigabitEthernet 0/0
```

```
[R2-GigabitEthernet0/0]ip address 172.16.1.254 24
[R2-GigabitEthernet0/0]quit
[R2]interface GigabitEthernet 0/1
[R2-GigabitEthernet0/1]ip address 20.20.20.10 24
[R2-GigabitEthernet0/1]quit
[R2]interface GigabitEthernet 0/2
[R2-GigabitEthernet0/2]ip address 40.40.40.10 24
[R2-GigabitEthernet0/2]quit
```

③ 配置路由器 R3 的接口：

```
<H3C>sys
System View: return to User View with Ctrl+Z.
[H3C]sysname R3
[R3]interface GigabitEthernet 0/0
[R3-GigabitEthernet0/0]ip address 10.10.10.254 24
[R3-GigabitEthernet0/0]quit
[R3]interface GigabitEthernet 0/1
[R3-GigabitEthernet0/1]ip address 40.40.40.1 24
[R3-GigabitEthernet0/1]quit
[R3]interface GigabitEthernet 0/2
[R3-GigabitEthernet0/2]ip address 30.30.30.10 24
[R3-GigabitEthernet0/2]quit
```

2）部署单区域 OSPF 网络

首先创建并运行 OSPF，然后创建区域并进入 OSPF 区域视图，指定运行 OSPF 协议的接口和接口所属的区域。

① 路由器 R1 的配置：

```
[R1]ospf 1
[R1-ospf-1]area 0
[R1-ospf-1-area-0.0.0.0]network 192.168.1.0 0.0.0.255
[R1-ospf-1-area-0.0.0.0]network 20.20.20.0 0.0.0.255
[R1-ospf-1-area-0.0.0.0]network 30.30.30.0 0.0.0.255
[R1-ospf-1-area-0.0.0.0]quit
```

② 路由器 R2 的配置：

```
[R2]ospf 1
[R2-ospf-1]area 0
[R2-ospf-1-area-0.0.0.0]network 172.16.1.0 0.0.0.255
[R2-ospf-1-area-0.0.0.0]network 20.20.20.0 0.0.0.255
[R2-ospf-1-area-0.0.0.0]network 40.40.40.0 0.0.0.255
[R2-ospf-1-area-0.0.0.0]quit
[R2-ospf-1]quit
```

③ 路由器 R3 的配置：

```
[R3]ospf 1
[R3-ospf-1]area 0
[R3-ospf-1-area-0.0.0.0]network 10.10.10.0 0.0.0.255
[R3-ospf-1-area-0.0.0.0]network 40.40.40.0 0.0.0.255
```

```
[R3-ospf-1-area-0.0.0.0]network 30.30.30.0 0.0.0.255
[R3-ospf-1-area-0.0.0.0]quit
[R3-ospf-1]quit
```

  3）在路由器上配置 SSH 登录

  ① 路由器 R1 的配置：

```
<R1>sys
System View: return to User View with Ctrl+Z.
[R1]public-key local create rsa
The range of public key modulus is (512 ~ 2048).
If the key modulus is greater than 512, it will take a few minutes.
Press CTRL+C to abort.
Input the modulus length [default = 1024]:2048
Generating Keys...
.
Created the key pair successfully.
[R1]ssh server enable
[R1]local-user admin
New local user added.
[R1-luser-manage-admin]password simple a123456789
[R1-luser-manage-admin]service-type ssh
[R1-luser-manage-admin]authorization-attribute user-role 3
[R1-luser-manage-admin]quit
[R1]user-interface vty 0 4
[R1-line-vty0-4]authentication-mode scheme
[R1-line-vty0-4]protocol inbound ssh
[R1-line-vty0-4]quit
[R1]save
```

  ② 路由器 R2 的配置：

```
<R2>sys
System View: return to User View with Ctrl+Z.
[R2]public-key local create rsa
The range of public key modulus is (512 ~ 2048).
If the key modulus is greater than 512, it will take a few minutes.
Press CTRL+C to abort.
Input the modulus length [default = 1024]:2048
Generating Keys...
.
Created the key pair successfully.
[R2]ssh server enable
[R2]local-user admin
New local user added.
[R2-luser-manage-admin]password simple a123456789
[R2-luser-manage-admin]service-type ssh
[R2-luser-manage-admin]authorization-attribute user-role 3
[R2-luser-manage-admin]quit
```

```
[R2]user-interface vty 0 4
[R2-line-vty0-4]authentication-mode scheme
[R2-line-vty0-4]protocol inbound ssh
[R2-line-vty0-4]quit
[R2]save
```

4）在网管计算机上安装模块

在网管计算机联网状态下安装模块 paramiko

```
//获取 pip 安装脚本
[root@manage ~]#curl "https://bootstrap.pypa.io/get-pip.py" -o "get-pip.py"
[root@manage ~]# python get-pip.py           //安装 pip 工具
[root@manage ~]# pip install paramiko        //通过 pip 安装 Python 第三方模块 paramiko
```

5）编写 Python 修改密码脚本

编写 Python 脚本 changepassword.py，实现修改路由器 R1、R2 和 R3 的管理密码。

```
[root@manage ~]# vi changepassword.py
    #导入 paramiko、time、getpass 模块
    #!/usr/bin/python
import paramiko
import time
import getpass
    #通过 input()函数获取用户输入的 SSH 用户名并将其赋值给 username
username = input('Username:')
    #通过 getpass 模块中的 getpass()函数获取用户输入的字符串，并将其作为密码并赋值给 password
password = getpass.getpass('Password:')
    #定义需要登录的设备列表
devices = ["192.168.1.254", "172.16.1.254", "10.10.10.254"]
    #遍历设备列表，并依次登录设备执行命令
for device in devices:
    ip = str(device)
    ssh_client = paramiko.SSHClient()        #创建一个 SSHClient 实例
    #设置自动添加主机密钥策略
    ssh_client.set_missing_host_key_policy(paramiko.AutoAddPolicy())
    ssh_client.connect(hostname=ip, username=username, password=password)
    command = ssh_client.invoke_shell()      #在交互式 shell 中执行命令
    #调度交换机在命令行中执行命令
command.send("system-view" + "\n")
command.send("local-user admin 0" + "\n")
command.send("password simple b123456789" + "\n")
    #更改管理密码完成后，返回用户视图并保存配置信息
command.send("return" + "\n")
command.send("save" + "\n")
command.send("Y" + "\n")
command.send("\n")
    #暂停 2s，并将命令执行过程赋值给 output 对象，通过 print (output.decode())语句回显内容
time.sleep(2)
```

```
output=command.recv(65535)
Print(output.decode())
    #退出 SSH
ssh_client.close()
```

6）编写 Python 备份脚本

在网管计算机上创建备份交换机运行配置的脚本 backup.py：

```
[root@manage ~]# vi backup.py
    #导入 paramiko、time、datetime 模块
    #!/usr/bin/python
import paramiko
import time
from datetime import datetime
    #设置SSH用户名和密码
username ="admin"
password ="234567"
    #定义需要登录的设备列表
devices = ["192.168.1.254", "172.16.1.254", "10.10.10.254"]
    #遍历设备列表，并依次登录设备执行命令
for device in devices:
    ip = str(device)
    ssh_client = paramiko.SSHClient()      #创建一个SSHClient实例
    #设置自动添加主机密钥策略
    ssh_client.set_missing_host_key_policy(paramiko.AutoAddPolicy())
    ssh_client.connect(hostname=ip, username=username, password=password)
    command = ssh_client.invoke_shell()      #在交互式shell中执行命令
    #提示 SSH 登录成功
    print "ssh "+ ip +"successfully"
    #设置回显内容不分屏显示
    command.send("screen-length disable " +"\n")
    #获取交换机运行配置
    output=(command.send("display current-configuration" +"\n"))
    #程序暂停2s
    time.sleep(2)
    #读取当前时间
    now=datetime.now()
    #打开备份文件
    #将备份文件保存在 backup 目录中
    backup=open("/root/backup/"+str(now.year)+"-"+str(now.month)+"-"+str(now.day)+"-"+ip+".txt","a+")
    #提示正在备份
    print "backuping"
    #将查询运行配置的回显内容，并将其赋值给 recv 对象
    recv=command.recv(65535)
    #将回显内容写入 backup 对象，相当于写入备份文件
    backup.write(recv)
```

```
#关闭打开的文件
backup.close()
#结束，断开 SSH 连接
ssh_client.close()
```

7）配置计划任务

配置计划任务实现每天凌晨 1 点自动执行脚本进行备份：

```
[root@manage ~]# vi /etc/crontab
    #在文件末尾添加以下代码，保存并退出
00 1 * * * root python /root/backup.py

[root@manage ~]# mkdir /root/backup              #创建一个名为 backup 的目录
[root@manage ~]# systemctl restart crond         #重启 crond 服务，以使设置生效
[root@manage ~]# systemctl enable crond          #在系统重启后自动启动定时任务
```

8）配置各计算机的 IP 地址

配置 PC2 的 IP 地址，如图 13-2 所示。

图 13-2　配置 PC2 的 IP 地址

配置 PC3 的 IP 地址，如图 13-3 所示。

图 13-3　配置 PC3 的 IP 地址

13.3.3 项目验证

1）执行 changepassword.py

执行 changepassword.py，查看回显内容：

```
[root@manage ~]# ./changepassword.py
Username:admin
Password:

User last login information:
Access Type: SSH
IP-Address : 192.168.1.1 ssh
Time       : 2023-04-29 11:31:45-08:00

<R1>system-view
Enter system view, return user view with Ctrl+Z.
[R1]local-user admin
[R1-luser-manage-admin]password simple b123456789
[R1-luser-manage-admin]return
<R1>save

The current configuration will be written to the device.
Are you sure to continue? (y/n)[n]:Y
It will take several minutes to save configuration file, please wait...

User last login information:

Access Type: SSH
IP-Address : 192.168.1.1 ssh
Time       : 2023-04-29 11:32:40-08:00

<R2>system-view
Enter system view, return user view with Ctrl+Z.
[R2]local-user admin
[R2-luser-manage-admin]password simple b123456789
[R2-luser-manage-admin]return
<R2>save
The current configuration will be written to the device.
Are you sure to continue? (y/n)[n]:Y
It will take several minutes to save configuration file, please wait...

User last login information:

Access Type: SSH
IP-Address : 192.168.1.1 ssh
Time       : 2023-04-29 11:34:41-08:00
```

```
<R3>system-view
Enter system view, return user view with Ctrl+Z.
[R3]local-user admin
[R3-luser-manage-admin]password simple b123456789
[R3-luser-manage-admin]return
<R3>save
The current configuration will be written to the device.
Are you sure to continue? (y/n)[n]:Y
It will take several minutes to save configuration file, please wait...
```

2）执行计划任务后查看备份文件

查看/root/backup 目录下的文件：

```
[root@manage ~]# cd /root/backup
[root@manage backup]# ls
2023-4-28-10.10.10.254.txt 2023-4-28-172.16.1.254.txt
2023-4-28-192.168.1.254.txt
```

```
[root@manage backup]# ll
total 12
-rw-r--r--. 1  root root 1786 Feb 28  1:00 2023-4-28-10.10.10.254.txt
-rw-r--r--. 1  root root 1809 Feb 28  1:00 2023-4-28-172.16.1.254.txt
-rw-r--r--. 1  root root 1762 Feb 28  1:00 2023-4-28-192.168.1.254.txt
```

查看详细内容：

```
[root@manage backup]# cat 2023-4-28-10.10.10.254.txt

User last login information:

Access Type: SSH
IP-Address : 192.168.1.1 ssh
Time       : 2023-04-29 10:32:24-08:00

<R3>screen-length disable
Info: The configuration takes effect on the current user terminal interface only.
<R3>display current-configuration
#version 7.1.064,Release 0821P11
#
sysname R3
#
snmp-agent local-engineid 800007DB03000000000000
snmp-agent
#
clock timezone China-Standard-Time minus 08:00:00
#
portal local-server load flash:/portalpage.zip
#
```

```
drop illegal-mac alarm
#
wlan ac-global carrier id other ac id 0
#
set cpu-usage threshold 80 restore 75
#
local-user admin password cipher %$%$YgN!G*Q*}0tjsqA"g~X(T{]!%$%$ local-user admin
privilege level 3
local-user admin service-type ssh
#
interface GigabitEthernet0/0
ip address 10.10.10.254 255.255.255.0
#
interface GigabitEthernet0/1
ip address 40.40.40.1 255.255.255.0
#
interface GigabitEthernet0/2
ip address 30.30.30.10 255.255.255.0
#
interface NULL0
#
ospf 1
area 0.0.0.0
network 10.10.10.0 0.0.0.255
network 30.30.30.0 0.0.0.255
network 40.40.40.0 0.0.0.255
#
stelnet server enable
#
line con 0
 user-role network-admin
 set authentication password hash $h$6$FU+Wk+yu07/IIL5f$5kSmN3+
kS1uekMpOGRyicknUalfr2j27wvTepXksqYMp1gqoKdvHDV+mJbknaAUTtBLoA//
R1AcC1BPi96EvqA==
#
line vty 0 4
 authentication-mode scheme
 user-role network-operator
 protocol inbound ssh
#local-user admin class manage
 password hash $h$6$w6LXVB7AEycxLblU$j8i+
bux12pQsknACOthj0xQb0FzcLhCaMGCDyEhztUPQPml8pB5PynnpwYyTOxJ4LwCQUS3b
tL/wfk1VyFeTQw==
 service-type ssh
 authorization-attribute user-role 3
```

```
authorization-attribute user-role network-operator
#
return
```

通过上面的程序可以发现，已备份路由器的整个配置信息。

13.4 项目相关知识

自动化运维知识

1）Python 模块

图 13-4 所示为 Python 模块。

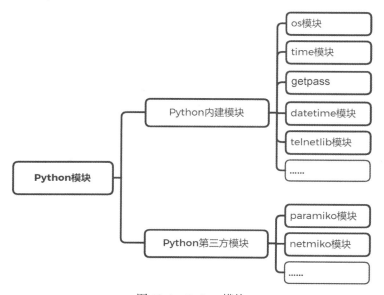

图 13-4　Python 模块

在 Python 中，模块可以被通俗地理解为独立保存的脚本。它可以通过 import module-name 语句来导入，其中 module-name 代表模块的名称。模块分为 Python 内建模块和 Python 第三方模块。其中，Python 内建模块可以直接通过 import module-name 语句来导入；Python 第三方模块可以先通过 pip install module-name 终端命令进行安装，再通过 import module-name 语句来导入。在网络运维中，常用的 Python 内建模块有 os、time、getpass、datetime 和 telnetlib 等，常用的 Python 第三方模块有 paramiko 和 netmiko 等。

2）网络运维常见的 Python 脚本

① 通过 getpass 模块提示用户输入密码并将用户输入的密码赋值给 a 对象：

```
import getpass
a=getpass.getpass('please input password:')
```

② 通过 time 模块暂停执行程序 60s：

```
import time
time.sleep(60)
```

③ 通过 datetime 模块将当前时间赋值给变量 a，并以"日-月-年 时:分"的形式回显出来：
```
import datetime
a = datetime.datetime.now()
print(a.day, "-", a.month, "-", a.year, a.hour, ":", a.minute)
```

　　getpass 模块是 Python 内建模块之一，主要用于提供 Python 的交互式功能，在网络运维中可以用于提示用户输入密码。通过 getpass 模块输入的密码是不可见的，安全性相对较高。

　　time 和 datetime 模块是 Python 内建模块之一，它们在 Python 中主要提供与时间相关的功能。time 模块可以在网络运维中提供时间戳、格式化时间等功能。datetime 模块重新封装了 time 模块，能够提供更多功能，如日期、时区等。

④ 通过 telnetlib 模块连接 IP 地址为 192.168.1.254 的 H3C 网络设备并执行 system-view 命令进入系统视图。其中，telnet 用户名为 admin，密码为 H3cH3c123。
```
import telnetlib
ip = "192.168.1.254"
user = "admin"
password = "H3cH3c123"
tn = telnetlib.Telnet(ip)
tn.read_until(b"Username:")                           #在字符串前加上 b 来将其转换为字节类型
tn.write(user.encode("utf-8") + b"\n")                #将用户名和密码转换为字节类型
tn.read_until(b"Password:")
tn.write(password.encode("utf-8") + b"\n")
tn.write(b"system-view" + b"\n")
```

　　telnetlib 模块主要用于支持 Python 通过 Telnet 协议远程连接设备，是 Python 内建模块，可以直接导入并使用，无须额外安装。然而，telnetlib 模块在数据传输过程中存在一些安全问题（如不支持密文传输），因此不太建议在生产网络中使用。

⑤ 通过 paramiko 模块连接 IP 地址为 192.168.1.254 的 H3C 网络设备并执行 system-view 命令进入系统视图。其中，SSH 用户名为 admin，密码为 H3cH3c1234。
```
import paramiko
username ="admin"
password ="H3cH3c1234"
ip="192.168.1.254"
ssh_client=paramiko.SSHClient()
ssh_client.set_missing_host_key_policy(paramiko.AutoAddPolicy())
ssh_client.connect(hostname=ip,username=username,password=password)
command=ssh_client.invoke_shell()
time.sleep(1)    #添加一个适当的延迟
output = command.recv(65535)   #接收一些输出
print(output.decode("utf-8"))  #打印输出
command.send("system-view" +"\n")
```

　　在 Python 中，安装并导入 paramiko 模块后可以通过代码实现 SSH 远程登录设备。具有同样效果的模块还有 netmiko。netmiko 模块主要在 paramiko 模块的基础上进行了优化，如增加厂商支持、增加命令补全功能等。

⑥ 调用 open()相关函数，以读写模式打开名为 backup.txt 文件，写入"abcd"内容后，

再将其读取出来：

```
#打开文件并以追加模式写入内容
with open('backup.txt', 'a+') as file:
    file.write('abcd')
    file.seek(0)    #将文件指针移动到文件的开始位置
    content = file.read()
print(content)    #输出读取到的内容
```

在上述代码中，首先使用 a+模式打开文件，并将内容"abcd"写入文件；然后使用 seek(0)将文件指针移动到文件的开始位置，这样才能读取到刚刚写入的内容；最后将读取到的内容打印出来。

注意：在使用 a+模式打开文件时，需要使用 with 语句，这样在退出 with 代码块时会正确关闭文件。另外，在使用 a+模式打开文件时，写入的内容会被追加到文件的末尾。如果只想写入新的内容而不是追加内容，那么可以使用 w 或 w+模式打开文件。

在日常的网络运维中，网络工程师需要使用文本文件来辅助工作，如使用批量配置网络设备的命令模板文件，存放所有网络设备的 IP 地址，以及备份网络设备运行配置信息命令 display current-configuration 输出的结果等。Python 内建模块 os 可以实现以上功能。os 模块中常用的函数有 open()。open()函数使用的代码格式一般为 open('filename','type')，其中 filename 代表文件名，type 代表文件的读写模式，可以为 r（只读）、w（写入）、a（追加）、r+（读写）、w+（覆盖读写）等。

13.5 项目小结

本项目通过批量自动更改交换机管理密码和自动备份的需求，展示了 Python 在网络自动化运维领域的具体应用，通过项目背景、项目需求分析、项目规划设计等环节，将项目实施部分拆分为多个子任务，符合工程项目实施的一般规律。

通过对本项目的学习，读者应该对项目实施流程有一定的了解，能够掌握 Python 在网络自动化运维领域的原理，并且能够熟练运用 Python 代码进行自动化和批量的运维操作。

素质拓展：密码技术 护网安全

今天，密码技术已经成为保障网络与信息安全的重要支撑，在确认真伪、鉴别来源方面的作用越来越凸显。近年来，信息化渗透到生产生活各个角落。老百姓利用网络数据来创造财富、通过计算机来鉴别凭证真伪、靠智能终端安排日程，信息安全成了每个人面临的现实问题。

事实上，在信息化环境中，我们每次下载验证代码的来源、每次支付确保数据的完整都需要密码技术支撑。

然而，与信息技术高速发展的需求相比，我国的密码产品无论是数量还是应用适应能力都存在一些不足。例如，许多老百姓在上网时使用的密码产品、密钥管理系统，往往是由国外的信息系统开发的，这给网络信息安全保障带来一些不确定性。

密码技术理论和工程联系紧密。只有产学研用协同努力，提高产品规模，增强产品适应能力，才能在实际应用中提升并优化密码技术。

增值服务

网络自动化运维极大地提升了网络增值服务。例如，可以提升服务的可用性。自动化运维能够实时监控网络设备的状态，及时发现并解决故障，确保服务的连续性和可用性；可以加速服务部署。传统的网络服务部署可能需要人工配置多个设备，而自动化运维能够通过脚本和工具完成自动配置，大大缩短了部署时间。自动化运维可以优化服务性能，通过对网络流量的监控和分析，能够发现并解决性能瓶颈，提升网络服务的整体性能；可以提升客户体验，快速响应客户需求，提供个性化的服务，从而提升客户满意度；可以降低运维成本，减少对人力干预的依赖，使得企业将更多的资源投入到创新和服务中；可以提供数据洞察，智能收集大量数据并将其转化为有价值的信息，为企业提供关于用户行为、服务性能等方面的洞察，为决策提供依据。作为网络管理员，要与时俱进，及时为客户提供新技术、新工艺、新方法。

13.6 课后实训

项目内容：公司网络拓扑规划如图 13-5 所示，设备规划表如表 13-3 所示。

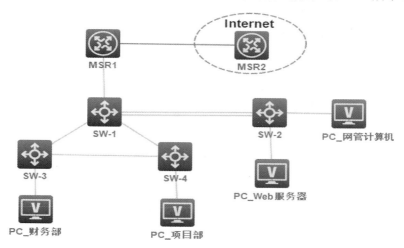

图 13-5　公司网络拓扑规划

表 13-3 设备规划表

所属区域	设备类型	型号	设备名称
园区网	路由器	MSR36-20	MSR2
核心机房	路由器	MSR36-20	MSR1
核心机房	三层交换机	S5820	SW-1
核心机房	三层交换机	S5820	SW-2
项目部	二层交换机	S3600	SW-3
财务部	二层交换机	S3600	SW-4
服务器群	网管计算机	PC	PC_网管计算机

公司网络内的所有路由器和交换机均已实现远程登录。其中，设备管理地址网段规划为 192.168.100.0/24，管理用户名为 admin，密码为 A123456，VTY 认证方式为 scheme；连接终端使用的网段为 192.168.1.0/24、192.168.2.0/24 和 192.168.3.0/24；出口 IP 地址为 202.111.1.1/24。本项目具体涉及的工作任务如下。

任务 1：自动修改网络设备的管理密码，主要通过在网管计算机上编写 Python 脚本来实现。

任务 2：定期自动备份网络设备配置，主要需要运用 Python 自动化运维的相关知识在网管计算机上编写 Python 脚本，读取网络设备的运行配置，并以"年-月-日-IP.txt"文件命名格式保存到系统~/backup 目录下。同时，需要配置系统的计划任务程序，使其每天凌晨 1 点自动执行一次。